The ArcGIS® Imagery Book

New View. New Vision.

Esri Press
Redlands, California

Esri Press, 380 New York Street, Redlands, California 92373-8100
Copyright © 2016 Esri
All rights reserved.

Printed in the United States of America
19 18 17 16 15   1 2 3 4 5 6 7 8 9 10

@esri.com, 3D Analyst, ACORN, Address Coder, ADF, AML, ArcAtlas, ArcCAD, ArcCatalog, ArcCOGO, ArcData, ArcDoc, ArcEdit, ArcEditor, ArcEurope, ArcExplorer, ArcExpress, ArcGIS, arcgis.com, ArcGlobe, ArcGrid, ArcIMS, ARC/INFO, ArcInfo, ArcInfo Librarian, ArcLessons, ArcLocation, ArcLogistics, ArcMap, ArcNetwork, *ArcNews*, ArcObjects, ArcOpen, ArcPad, ArcPlot, ArcPress, ArcPy, ArcReader, ArcScan, ArcScene, ArcSchool, ArcScripts, ArcSDE, ArcSdl, ArcSketch, ArcStorm, ArcSurvey, ArcTIN, ArcToolbox, ArcTools, ArcUSA, *ArcUser*, ArcView, ArcVoyager, *ArcWatch*, ArcWeb, ArcWorld, ArcXML, Atlas GIS, AtlasWare, Avenue, BAO, Business Analyst, Business Analyst Online, BusinessMAP, CityEngine, CommunityInfo, Database Integrator, DBI Kit, EDN, Esri, esri.com, Esri—Team GIS, *Esri—The GIS Company*, Esri—The GIS People, Esri—The GIS Software Leader, FormEdit, GeoCollector, Geographic Design System, Geography Matters, Geography Network, geographynetwork.com, Geoloqi, Geotrigger, GIS by Esri, gis.com, GISData Server, GIS Day, gisday.com, GIS for Everyone, JTX, MapIt, Maplex, MapObjects, MapStudio, ModelBuilder, MOLE, MPS—Atlas, PLTS, Rent-a-Tech, SDE, SML, Sourcebook•America, SpatiaLABS, Spatial Database Engine, StreetMap, Tapestry, the ARC/INFO logo, the ArcGIS Explorer logo, the ArcGIS logo, the ArcPad logo, the Esri globe logo, the Esri Press logo, The Geographic Advantage, The Geographic Approach, the GIS Day logo, the MapIt logo, The World's Leading Desktop GIS, *Water Writes*, and Your Personal Geographic Information System are trademarks, service marks, or registered marks of Esri in the United States, the European Community, or certain other jurisdictions. CityEngine is a registered trademark of Procedural AG and is distributed under license by Esri. Other companies and products or services mentioned herein may be trademarks, service marks, or registered marks of their respective mark owners.

Ask for Esri Press titles at your local bookstore or order by calling 800-447-9778, or shop online at esri.com/esripress. Outside the United States, contact your local Esri distributor or shop online at eurospanbookstore.com/esri.

Esri Press titles are distributed to the trade by the following:

In North America:
Ingram Publisher Services
Toll-free telephone: 800-648-3104 Toll-free fax: 800-838-1149
Email: customerservice@ingrampublisherservices.com

In the United Kingdom, Europe, Middle East and Africa, Asia, and Australia:
Eurospan Group, 3 Henrietta Street, London WC2E 8LU, United Kingdom
Telephone: 44(0) 1767 604972 Fax: 44(0) 1767 601640
Email: eurospan@turpin-distribution.com

All images courtesy of Esri except as noted.

On the front cover: *This digital portrait of Earth was inspired by the Apollo-era pictures of the "big blue marble" Earth from space. To create it, data scientists at Goddard Space Flight Center's Laboratory for Atmospheres combined data from a Geostationary Operational Environmental Satellite (GOES), the Sea-viewing Wide Field-of-view Sensor (SeaWiFS), and the Polar Orbiting Environmental Satellites (POES) with a USGS elevation model of Earth's topography. Stunningly detailed, the planet's western hemisphere is cast so that heavy vegetation is green and sparse vegetation is yellow, while the heights of mountains and depths of valleys have been exaggerated by 50 times to make vertical relief visible. Hurricane Linda is the dramatic storm off North America's west coast.*

On the back cover: *Astronaut William Anders and the rest of the Apollo 8 crew became the first humans to leave Earth orbit, entering lunar orbit on Christmas Eve, 1968. This iconic image captured by Anders with a Hasselblad camera stunned the world when it was published.*

Table of Contents

Foreword

A special relationship has always existed between GIS and remote sensing, and it goes back to the very beginning of our modern information technology. In the 1960s and 1970s, computer systems for GIS were big, expensive, and very slow mainframes using punched cards, but nearly all the foundation data layers in these early systems came either directly or indirectly from imagery. Right from the start, GIS and remote sensing were complementary, like two sides of the same coin. They were coevolving together.

In 1972, a revolution happened with the launch of Landsat—the first commercial earth observation imaging satellite. It continuously orbited the earth and captured a new image of the same spot about every 16 days. Because it was so high up, it gave us an entirely different picture of our planet and its patterns. It provided not only a new view; it gave us a new vision of the possibility of what GIS could become. And it started a revolution in commercial earth observation that continues today and is exploding now with hundreds—and soon thousands—of smaller satellites, microsatellites, video cameras from space, high-altitude drones, and more.

So where are GIS and remote sensing—these two close allies for more than 50 years—going next?

For one thing, there's a big emphasis now on simplicity and speed. It's clear that the future belongs to the simple and quick. We're seeing that modern technology is harnessing this amazing array of globally distributed sensors into what is popularly referred to as the Internet of Things, a vast collection of dynamic, live information streams that are feeding into and becoming the heart of web GIS. Plus, this network operates in real time,

giving us access to what we might call the "Internet of All My Things"—and all on our own devices through a new geoinformation model.

Although the technology powering this concept is advanced, we comprehend it in practice because we understand pictures. Einstein famously said, "If I can't see it, I can't understand it." We know something when we can see it.

And now, all of these rapidly changing developments combining imagery and spatial analyses are opening up new chapters in the history of GIS, as society is awakening to the power of geography and the intuitive understanding that imagery helps us "see" in all its forms.

We like to say that the map of the future is an intelligent image.

Lawrie Jordan is Esri's Director of Imagery and Remote Sensing. He is a pioneer in the field of image processing and remote sensing.

 Video: The map of the future is an intelligent image

How this book works

The purpose of this book is to show you—the GIS professional, app developer, web designer, or virtually any other type of technologist—how to become a GIS and imagery ace. Or put another way, to become someone who is a smarter, more skillful, and more powerful applier of image data within a GIS. Imagery is suddenly a big deal, and those who are adept at finding it, analyzing it, and understanding what it actually means are going to be in demand in the years ahead.

Audience

There are several potential audiences for this book. The first is the worldwide professional GIS and mapping community, the people who work with maps and geospatial data every day, in particular those who wish to do more with imagery in their GIS applications. If you're a data scientist, cartographer, part of a government agency staff, urban planner, or other GIS professional, you already may be leveraging the web and pushing geographic information out to the public. You may already instinctively recognize the inherent value of imagery as an amazing data capture technology that integrates well with traditional vector-based geospatial data.

Another audience includes people new to GIS with an interest in things you can do with imagery—people like amateur drone pilots flying missions to map school campuses, real estate developers planning redevelopment projects, or citizen scientists and bloggers reporting about climate change who are perhaps coming to GIS through an interest in imagery.

Finally, this book will be of interest to people who just love to explore the world and look at fascinating pictures of the earth. For these "armchair" geographers and the rest, this book and the electronic companion found at *TheArcGISImageryBook.com* offer a wealth of gorgeous and sometimes troubling images as well as links to powerful imagery-based web apps and maps that weave interesting stories about our planet. The only prerequisite to benefit from this book is a desire to better understand your world through imagery and mapping, plus a roll-up-your-sleeves attitude.

Learn by doing

This is a book that you do as well as read, and all you really need is a personal computer with web access. The adventure starts when you engage yourself in the process by opening the links, exploring maps and apps that others have made, and then doing the lessons to create your own maps and apps. These resources (over 200 maps, apps, videos, and images in all) are hyperlinked at *TheArcGISImageryBook.com*.

This is a book about applying imagery in ArcGIS, the web GIS platform, and is the second in a series of Big Idea titles. If you're new to GIS, you may want to check out the first in the series, *The ArcGIS Book: 10 Big Ideas About Applying Geography to Your World*. While this volume is designed as a stand-alone work, many readers will also find the original book of interest.

A word on interactive content

When you see references in the text to interactive web content, you will need to be in either the web version or the interactive PDF version of this book at www.thearcgisimagerybook.com.

These cloud patterns cast eerie shadows on the landscape of southern Egypt. The clouds appear red and the desert below hazy blue in this infrared rendition. The black circles are center-pivot irrigated farms.

Imagery Is Visible Intelligence
A geographic Rosetta stone

Geographic information system (GIS) technology is both intuitive and cognitive. It combines powerful visualization and mapping with strong analytic and modeling tools. Remotely sensed earth observation—generally referred to in GIS circles simply as *imagery*—is the definitive visual reference at the heart of GIS. It provides the key, the geographic Rosetta stone, that unlocks the mysteries of how the planet operates and brings it to life. When we see photos of Earth taken from above, we understand immediately what GIS is all about.

Imagery deepens understanding
Seeing is not only believing, but also perceiving

The story of imagery as an earth observation tool begins with photography, and in the early part of the twentieth century, photography underwent extraordinary changes and social adoption. Photos not only offered humanity a new, accessible kind of visual representation—they also offered a change in perspective. The use of color photography grew. Motion pictures and television evolved into what we know today. And humans took to the sky flying in airplanes, which, for the first time, enabled us to take pictures of the earth from above. It was a time of transformation in mapping and observation, providing an entirely new way of seeing the world.

World War II: Reconnaissance and intelligence gathering

During World War II, major advances in the use of imagery for intelligence were developed. The Allied Forces began to use offset photographs of the same area of interest, combining them to generate stereo photo pairs for enhancing their intelligence gathering activities. In one of many intelligence exercises called Operation Crossbow, pilots flying in planes—modified so heavily for photo gathering that there was no room for weapons—captured thousands of photographs over enemy-held territory. These resulting collections required interpretation and analysis of hundreds of thousands of stereo-photographic pairs by intelligence analysts.

These 3D aerial photographs enabled analysts to identify precise locations of highly camouflaged rocket technology developed by Germany. This was key in compromising the rocket systems that were targeting Great Britain, thus saving thousands of lives and contributing to ending World War II. The BBC did an excellent documentary on this subject (*Operation Crossbow: How 3D glasses helped defeat Hitler*).

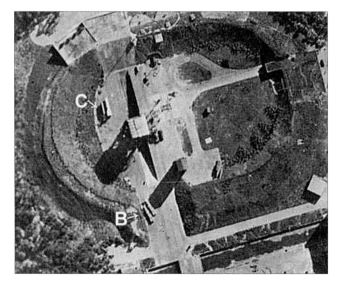

Stereoscopic imagery was instrumental in identifying the facilities of Nazi rocket programs. The photo above shows stereo glasses used for viewing offset photo pairs. This June 1943 photograph (left) was the first to reveal functional weapons. Two V2 rockets 40 feet long are seen lying horizontally at (B), but only in December was it realized that the structure at (C) was a prototype flying-bomb catapult.

1969: Dawn of extraterrestrial man

The first humans explore our moon

In the early 1960s, the majority of people would probably have said it was impossible for a human being to walk on the moon. But in July 1969, televised images transmitted to Earth from the moon showed Neil Armstrong and Buzz Aldrin bounding across the lunar surface, proving that moon walking was more than conceptually possible—it was happening right before our eyes. Seeing was believing.

When Armstrong, Aldrin, and the ensuing lunar astronauts pointed their cameras back at Earth, an unexpected benefit became apparent: humanity now had a completely new perspective about our home planet—heralding the adoption and use of earth imagery.

In December 1972, Apollo 17 astronauts captured this iconic photo of Earth from space—the famous "Blue Marble" photograph, offering humanity a new perspective about Planet Earth and our place in the universe.

Astronaut Buzz Aldrin, from the Apollo 11 mission, on the moon in July 1969. Photo by Astronaut Neil Armstrong (visible in Aldrin's face shield).

1972: The Landsat program
Providing the first satellite images covering Earth

In 1972, the same space technology that was developed to put humans on the moon led to the launch of the first Landsat satellite. The Landsat mission gave us extraordinary new kinds of views of our own planet. This was a breakthrough system and the first civilian-oriented, widely available satellite imagery that not only showed us what was visible on Earth—it also provided a view of invisible information, unlocking access to electromagnetic reflections of our world as well. We could see Earth in a whole new way.

This persistent earth observation program continues to this day along with hundreds of other satellites and remote sensing missions as well. Nation states and, more recently, private companies have also launched numerous missions to capture earth imagery, allowing us to continuously observe and monitor our planet.

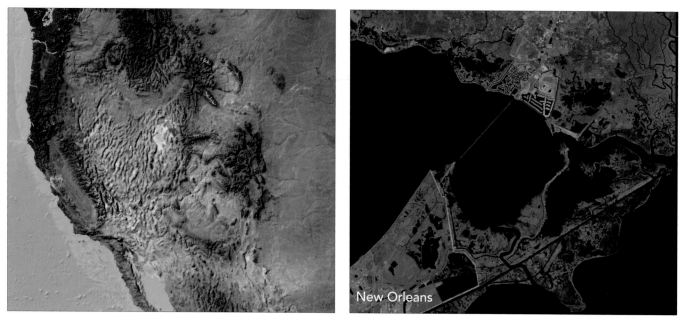

Landsat sensors have been continually generating and sharing pictures of the earth since the 1970s. Early on, scientists were excited by the valuable new perspectives being generated. Today, huge numbers of satellites image the earth thousands of times daily, creating a massive and virtual image catalog of our planet. Web GIS is tapping into these images to enable practitioners to address a broad array of questions and challenges that we face as Earth's stewards.

Show me my home!

2005: The human era of GIS begins

Little more than a decade ago, seemingly the whole world snapped awake to the power of imagery of the earth from above. We began by exploring a continuous, multiscale image map of the world provided online by Google and other mapping companies. A combination of satellite and aerial photography, these pictures of Earth helped us to experience the power of imagery, and people everywhere began to experience some of what GIS practitioners already knew. We immediately zoomed in on our neighborhoods and saw locational contexts for where we reside in the world. This emerging capability allowed us to see our local communities and neighborhoods through a marvelous new microscope. Eventually, naturally, we focused beyond that first local exploration to see anywhere in the world. What resulted was a whole new way to experience and think about the world.

These simple pictures captured people's imagination, providing whole new perspectives, and inspired new possibilities. Today, virtually anyone with Internet access can zero in on their own neighborhood to see their day-to-day world in entirely new ways. In addition, people everywhere truly appreciate the power of combining all kinds of map layers with imagery for a richer, more significant understanding.

Almost overnight, everyone with access to a computer became a GIS user.

Initially, we zoomed in on our homes and explored our neighborhoods through this new lens. This experience transformed how people everywhere began to more fully understand their place in the world. We immediately visited other places that we knew about. Today, we continue by traveling to faraway places we want to visit. Aerial photos provide a new context from the sky and have forever changed our human perspective. This map tour visits selected areas in several communities where ultra-high-resolution imagery is available.

Imagery expands your perspective
Seeing the visible, the invisible, the past, and into the future

Seeing is believing. Observing the world in colorful imagery is informative and immediate, delivering stark visual evidence and new insights. Imagery goes far beyond what our own eyes are capable of showing us—it also enables us to see our world in its present state. And it provides a means to look into the past as well as forecast the future, to perceive and understand Earth, its processes, and the effects and timelines of human activity. Amazingly, imagery even allows us to glimpse the invisible, to see visual representations of reflected energy across the entire electromagnetic spectrum, and to thus make more fully informed decisions about the critical issues facing Earth and all its life-forms.

Understanding seasonal climate patterns
Global imagery is collected continuously, enabling us to witness our world in action. By combining images from across the span of time, we can begin to visualize, animate, analyze, and understand Earth's cycles, where we come from, and where we are going.

Weather provides "breathing ranges of snow and ice" for the planet, delivering precious water that enables and sustains all living things. This image shows the seasonal weather cycles of precipitation across the North American continent.

Seeing beyond the visible
Imagery enables us to see beyond what our human eyes perceive, providing new scientific perspectives about Earth. Satellites have sensors that measure nonvisible information, such as infrared energy, across the electromagnetic energy spectrum that enables us to generate and analyze a multitude of new terrestrial views of our world.

This false color image over North Africa shows dry and wet areas of vegetation moisture by looking at the near infrared (band 5) and the shortwave infrared (band 6). Warmer colors reflect arid areas in the map. The striping pattern shows Landsat 8's scene footprints, illustrating its continuous orbit of Earth and revisiting every scene location roughly every 16 days.

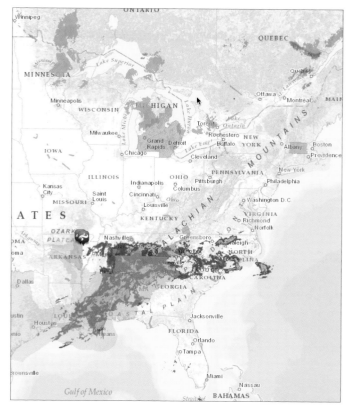

Forecasting and tracking daily weather

Advances in imaging and weather observations over the last decade have resulted in a dramatic increase in the accuracy and precision of weather forecasts. GIS integration of weather data for operations management has expanded to benefit farmers, emergency response teams, school districts, utilities, and many others. The sensors range from global weather satellites to ground-based local weather instruments that allow experts to monitor and forecast weather events like never before. The sensor network has become hyperlocal, allowing continuous forecasting of weather events in our communities. We can now access an accurate weather forecast for our neighborhood for the upcoming hour.

This radar-derived layer for the US mainland from AccuWeather shows precipitation in near real time. These near real-time weather observations, along with weather forecasting, are managed using image observations.

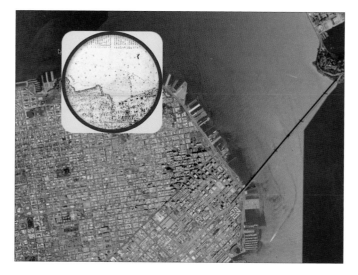

Beyond the apparent

Imagery enables you to peer into the past as well as combine historic views with current imagery. Imagery comes in a very simple format, allowing it to be easily overlaid with other maps and images into a kind of layered "virtual sandwich."

This unique spyglass app shows how the city of San Francisco expanded beyond the historic coastline settlement. With the city's location along the San Andreas Fault, expanding into the bay had its unique challenges requiring construction engineers to push piling down about 200 feet to bedrock.

Thought Leader: Jack Dangermond
GIS now includes a complete image processing system

Since the early days in our field, GIS professionals performed image processing and GIS in separate systems. For years, due to computing and storage limitations, these separate systems were necessary, and some users required both. Users would perform image processing in their separate imagery system to generate key data layers and then feed the results into their GIS. Even though both systems had geographic foundations, we took it for granted that these were separate systems and tasks—we even had separate people to run these systems. And each had their own communities and approaches. There was a general expectation and understanding that imaging systems and GIS were separate technologies but nonetheless closely linked.

From a practical point of view, however, this made little sense. Both technologies required a geospatial framework for their information sets. Both systems managed these datasets as geographic layers. Both provided layers that are georeferenced so they can be combined, mashed up, and overlaid with other layers. Of course one integrated system makes sense.

Only recently have GIS and imagery been brought together in one integrated and complete system—within ArcGIS. Part of this is the new image processing capabilities that have been added to ArcGIS. Big image data processing and management now work seamlessly with the continuously expanding web of earth observation data; drone missions; and new ground-, air-, and space-based sensors rapidly coming online. The immediate new trend is the migration of these sophisticated processing capabilities onto the massively scaleable cloud computing networks that enable ArcGIS to work with your big data collections.

Jack Dangermond is president and founder of Esri, the world leader in GIS software development.

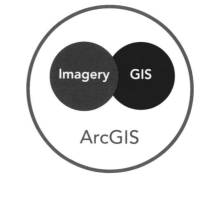

Jack Dangermond discusses the relevance of imagery to GIS

Imagery has so many uses
A range of applications

By now, it's apparent that imagery enables whole new perspectives and insights into your world and the issues you want to address. Imagery also has numerous advantages and capabilities.

Almost daily access to new information
Image collection is rapid and increasing. And access to imagery is increasingly becoming more responsive. Many satellites and sensors are already deployed with more coming all the time, collecting new data, adding to a continuous collection effort—a time series of observations about our planet. These image collections are enabling us to map, measure, and monitor virtually everything on or near the earth's surface. All of us can quite rapidly gather much of the data that we need for our work. Imagery has become our primary method for exploration when we "travel" to other planets and beyond. We send probes into space and receive returns primarily in the form of imagery that provides a continuous time series of information observations. And it enables us to derive new information in many interesting ways.

Looking back in time
The use of aerial imagery is still relatively young. While imagery only began to be used in the twentieth century, it is easy to compare observations for existing points in time that reside in our imagery collections. In addition, we can overlay imagery with historical maps, enabling us to compare the past with the present.

Imagery data collections are becoming richer every day
Imagery is creating an explosion of discovery. Many imagery initiatives are repetitive and growing, expanding and adding to image databases for our areas of interest. ArcGIS is scaling out, enabling the management of increasingly large, dynamically growing earth observations. This points to the immediacy of imagery and its capacity for easy integration, enabling all kinds of new applications and opportunities for use—things like before-and-after views for disaster response, rapid exploitation of newly collected imagery, image interpretation and classification, and the ability to derive intelligence. Over time, many of these techniques will grow in interesting new ways, enabling deeper learning about our communities, the problems and issues we face, and how we can use GIS to address these.

Imagery enables powerful analytic capabilities
Imagery and its general raster format enable rich analysis using ArcGIS. And, in turn, these enable more meaningful insights and perspectives about the problems we want to address.

Working together at last
Combining GIS and image processing provides synergy

Imagery in all of its variations uses one of the key common data formats in GIS—something called rasters. It is one of the most versatile GIS data formats. Virtually any data layer can be conveyed as a raster. This means that you can combine all kinds of data with your imagery, enabling integration and analytics.

Rasters provide a host of useful GIS data layers

Rasters, like any digital photo, provide a data model that covers a mapped area with a series of pixels or cells of equal size that are arranged into a series of rows and columns. They can be used to represent pictures as collections of pixels, surfaces such as elevation or proximity to selected features, all kinds of features themselves (in other words, points, lines, and areas), and time series information with many states for each time period.

Classified land cover and land use

Land cover around the western Mediterranean, from a global raster dataset from MDA of the predominant land characteristics at 30-meter resolution.

Distance to water

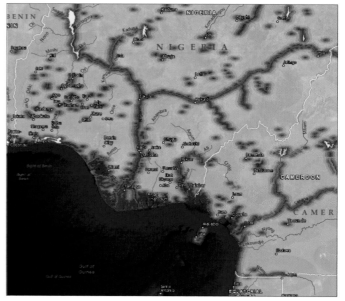

A proximity map showing distance from each cell or pixel to a reliable water source in a portion of West Africa. Water access is vital for humans as well as wildlife habitat. Streams are overlaid on the distance grid. Cells in the grid that are closest to water are darker blue. Colors change as the distance from water increases.

Three-dimensional scenes

Mont Blanc or Monte Bianco in the Alps between France and Italy. The app features a 3D tour of interesting sites from around the world.

Elevation expressed as shaded relief

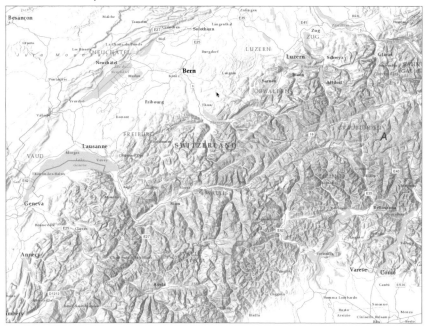

Global elevation displayed as shaded relief. This is part of a global elevation layer compiled from best available sources worldwide.

Oblique perspective photos

Oblique imagery provides a special perspective view of real-world features, presenting natural detail in 3D and enabling interpretation and reconnaissance.

Time series information

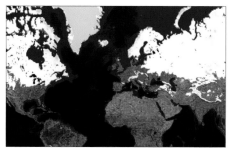

A snapshot of a time-enabled image map of monthly snowpack observations from the NASA Global Land Data Assimilation System (GLDAS). This map contains cumulative snowpack depths for each month from 2000 through 2015.

The Lena River is one of the largest rivers in the world. The Delta Reserve is the most extensive protected wilderness area in Russia. This refuge is a breeding ground for many species of Siberian wildlife.

Imagery is beautiful
Both informative and sublime

While imagery provides whole new perspectives, profoundly shaping our understanding, it's also clear that imagery provides exquisite views of our world—truly stunning and beautiful works of art. They astonish and amaze us, tapping into our emotions and the wonder of our world and new worlds we seek to discover and explore. It's no accident that the US Geological Survey maintains a collection of <u>Earth as Art</u>.

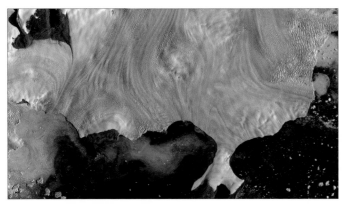

Along Greenland's western coast, a small field of glaciers surrounds Baffin Bay.

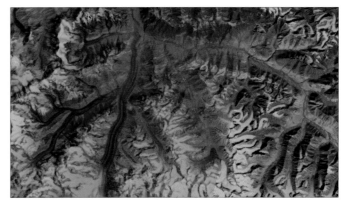

Soaring, snow-capped peaks and ridges of the eastern Himalaya Mountains create an irregular blue-on-red patchwork between major rivers in southwestern China.

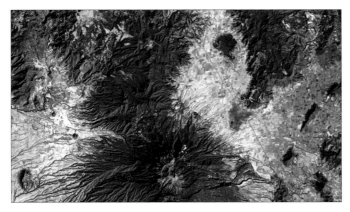

Snow-capped Colima Volcano, the most active volcano in Mexico, rises abruptly from the surrounding landscape in the state of Jalisco.

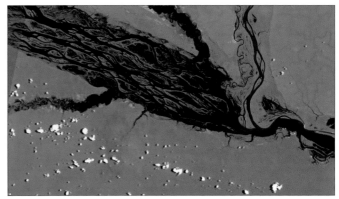

Fed by multiple waterways, Brazil's Negro River is the Amazon River's largest tributary. The mosaic of partially submerged islands visible in the channel disappears when rainy season downpours raise the water level.

Earth from space
The power of a single image

In the run-up to the Apollo moon landings, Apollo 8 was the first mission to put humans into lunar orbit. And on Christmas Eve 1968, coming around from the far side of the moon during their fourth orbit, Apollo 8 commander Frank Borman exclaimed, "Oh my God, look at that picture over there! Here's the Earth coming up! Wow, that is pretty!" Fellow astronaut Bill Anders grabbed his Hasselblad camera and shot this now-famous image of Earth rising above the moon.

In his book *Earthrise: How Man First Saw the Earth*, historian Robert Poole suggests that this single image marked the beginning of the environmental movement, saying that "it is possible to see that Earthrise marked the tipping point, the moment when the sense of the space age flipped from what it meant for space to what it means for Earth." The power of imagery can be neatly summed up in the story of this single photograph. Images can help us to better understand our planet, drive change, create connections—and in some cases even start a movement.

It's one of the most frequently reproduced and instantly recognizable photographs in history. The US Postal Service used the image on a stamp. Time magazine featured it on the cover. It was—and still is—"the most influential environmental photograph ever taken," according to acclaimed nature photographer Galen Rowell.

Mapping the solar system
An effort reflecting humanity's seeking spirit

Since the first moon shots, astronaut-photographers from the world's space agencies have also been turning their lenses *away* from Earth. GIS people, being the science fanatics they often are, have of course found ways to map planetary bodies other than our home planet. In 2015, NASA announced to the world that multispectral imagery taken from Mars-orbiting sensors had definitively ascertained the presence of moving water on Mars—a milestone not lost on the GIS and image analysis community.

3D visualization of the hyperspectral imagery data that changed our perception of the planet Mars.

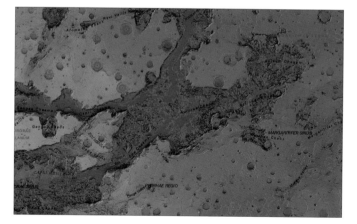

This map paints a picture of the dramatic Mars geography and all the surface missions that humans have carried out in exploration of the faraway red planet.

After a 3-billion-mile, nearly 10-year journey, on July 14, 2015, the New Horizons interplanetary space probe became the first spacecraft to explore the dwarf planet Pluto's moon Charon.

Quickstart

Connect with and deploy the ArcGIS platform

Now it's time get your hands on ArcGIS. If you're an existing user and already have an ArcGIS subscription (with publisher privileges), as well as ArcGIS Pro installed on your local machine, you have everything you need and can skip to the next page. If you don't yet have these two things, read on.

▸ **Get a Learn ArcGIS organization membership**

The majority of lessons in this book are carried out on the ArcGIS platform (in the cloud), and require membership (with Publisher privileges) in an ArcGIS organization. The Learn ArcGIS organization is available for students and others just getting started with ArcGIS. With your membership, you can immediately begin to use maps, explore data resources, and publish geographic information to the web. Go to the Learn ArcGIS organization and click the Sign Up Now link to activate a 60-day membership.

▸ **Install ArcGIS Pro**

ArcGIS Pro is a desktop application that you download and install on your local computer. It is licensed to you for 60 days through your membership in the Learn ArcGIS organization. Check the system requirements and then use the download button below to install the software on your local machine.

▸ **System requirements**

ArcGIS Pro is a 64-bit Windows application. To see if your computer will run ArcGIS Pro, click check the requirements.

Download ArcGIS Pro

The Learn ArcGIS organization is set up specifically for student use. You can join this organization even if you already have another ArcGIS account.

Learn ArcGIS Lesson

Get Started with Imagery

In this lesson, you'll explore Landsat imagery and some of its uses with the Esri Landsat app. You'll first go to the Sundarbans mangrove forest in Bangladesh, where you'll see the forest in color infrared and track vegetation health and land cover. Then, you'll find water in the Taklamakan Desert and discover submerged islands in the Maldives. After using 40 years of stockpiled Landsat imagery to track development of the Suez Canal over time, you'll be ready to explore the world on your own.

▶ **Overview**

Satellite imagery is an increasingly powerful tool for mapping and visualizing the world. No other method of imagery acquisition encompasses as much area in as little time. The longest-running satellite imagery program is Landsat, a joint initiative between two US government agencies. Its high-quality data appears in many wavelengths across the electromagnetic spectrum, emphasizing features otherwise invisible to the human eye and allowing a wide array of practical applications.

▶ **Build skills in these areas:**

- Navigating and exploring imagery
- Changing spectral bands to emphasize features
- Tracking changes over time
- Building your own band combination

▶ **What you need:**

- Estimated time: 30 minutes

Start Lesson

Esri.com/imagerybook/Chapter1_Lesson

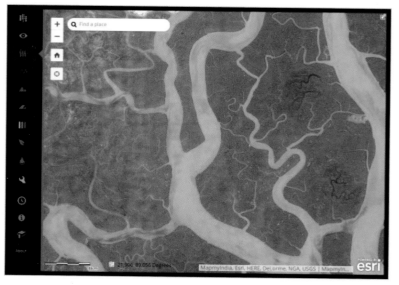

This app allows you to navigate the world with Landsat satellite imagery. Landsat takes images of the planet to reveal its secrets, from volcanic activity to urban sprawl. Landsat sees things on the electromagnetic spectrum, including what's invisible to the human eye. Different spectral bands yield insight about our precious and continually changing Earth.

02

The Nature of Remote Sensing
Information gathered from a distance

Remote sensing—the acquisition of information from a distance—has had a profound impact on human affairs in modern history. This image of British Beach (the WWII code name for one landing spot of the June 1944 Normandy invasion) taken from a specially equipped US Army F5, reveals rifle troops on the beach coming in from various large and small landing craft. Seven decades later—even as its application has expanded to unimaginable reaches—remote sensing remains the most significant of reconnaissance and earth observation technologies.

The view from above
The power of remote sensing

Humans have always sought the high vantage point above the landscape. Throughout history, whether from a treetop or a mountain peak or a rocky cliff, the view from above allowed our ancestors to answer important questions: Where is there water? Where is the best hunting ground? Where are my enemies? Aerial photography was first practiced by balloonist Gaspard-Félix Tournachon in 1858 over Paris. With the advent of both photography and practical airflight in the early twentieth century, the advantages of having the high ground led to a quantum shift forward and the field of remote sensing was born.

The technology came of age rapidly during World War I as a superior new military capability. From 1914 to 1918, aerial reconnaissance evolved from basically nothing to a rigorous, complex science. Many of the remote sensing procedures, methods, and terminology still in use today had their origins in this period. Throughout World War II the science and accuracy of remote sensing increased.

The next big evolutionary step came with spaceflight and digital photography. Satellite technology allowed the entire globe to be repeatedly imaged, and digital image management and transmission made these expanding volumes of images more useful and directly applicable. Today's diverse human endeavors require a steady flow of imagery, much of which finds its way onto the web within moments of capture.

The use of aerial photography rapidly matured during the First World War, as aircraft used for reconnaissance purposes were outfitted with cameras to record enemy movements and defenses. At the start of the conflict, the usefulness of aerial photography was not fully appreciated, with reconnaissance being accomplished by cartographers sketching out maps from the air.

Remote sensing capturing history

How remotely sensed images document stark truths

The first aerial photograph was taken in 1858, a century before the term "remote sensing" came into existence. Long before satellites and digital image capture became available, people were taking pictures of the earth's surface from afar, documenting many crucial moments in history for posterity.

One of the earliest aerial photographs to gain world renown is of the ruins of San Francisco, California, after the 1906 earthquake. It is a 160-degree panorama taken from a kite 2,000 feet (610 m) in the air above San Francisco that showed the entire city on a single 17-by-48-inch contact print made from a single piece of film. This image by commercial photographer George Lawrence documented the extensive fire damage across the city.

This Landsat 7 image of Manhattan on September 12, 2001, shows the extent of the toxic plume spreading over large portions of New York and New Jersey.

On October 14, 1962, American aerial photographs of Cuba revealed missile erectors, fuel tank trailers, and oxidizer tank trailers.

Many platforms, many applications
Sensor altitude plays a role in determing purpose

Modern imagery is captured from a broad range of altitudes starting from ground level to over 22,000 miles above Earth. The images that come from each altitude offer distinct advantages for each application. While not meant to be an exhaustive inventory, this diagram breaks down some of the most commonly used sensor altitudes:

Syncom

Geosynchronous ———————————————— 22,236 miles

Satellites that match Earth's rotation appear stationary in the sky to ground observers. While most commonly used for communications, geosynchronous orbiting satellites like the hyperspectral *GIFTS* imager are also useful for monitoring changing phenomena such as weather conditions. NASA's Syncom, launched in the early 1960s, was the first successful "high flyer."

Landsat 8

Sun synchronous ———————————————— 375–500 miles

Satellites in this orbit keep the angle of sunlight on the surface of the earth as consistent as possible, which means that scientists can compare images from the same season over several years, as with Landsat imagery. This is the bread-and-butter zone for earth observing sensors.

Helios

Atmospheric satellite ———————————————— 100,000 feet

Also known as pseudo-satellites, these unmanned vehicles skim the highest edges of detectable atmosphere. NASA's experimental Helios craft measured solar flares before crashing in the Pacific Ocean near Kauai.

SR71 Blackbird

Jet aircraft

90,000–30,000 feet

Jet aircraft flying at 30,000 feet and higher can be flown over disaster areas in a very short time, making them a good platform for certain types of optical and multispectral image applications.

Cessna

General aviation aircraft

100–10,000 feet

Small aircraft able to fly at low speed and low altitude have long been the sweet spot for high-quality aerial and orthophotography. From Cessnas to ultralights to helicopters, these are the workhorses of urban optical imagery.

Helicopter

Ultralight

Drones

100–500 feet

Drones are the new kids on the block. Their ability to fly low, hover, and be remotely controlled offer attractive advantages for aerial photography, with resolution down to sub-1 inch. Military UAVs can be either smaller drones or actual airplanes.

US Navy Silver Fox

3DR Solo private drone

Handheld spectrometer

Ground based/handheld

Ground level

Increasingly, imagery taken at ground level is finding its way into GIS workflows. Things like Google Street View, HERE streel-level imagery, and Mapillary; handheld multispectral imagers; and other terrestrial sensors are finding applications in areas like pipelines, security, tourism, real estate, natural resources, and entertainment.

Smartphone

Street-level mapping car

Imagery application trends

Remote sensing is already part of many industries

As the authoritative record of changing conditions on the ground, remote sensing imagery has a broad array of applications in traditional terrestrial human activities that involve the management of land. As such, industries like forestry, agriculture, mining, and exploration were among the early adopters of remote sensing, funding its growth.

Precision agriculture

Information gathered during harvest, including yield at any given location, helps growers track their results and provides valuable input for calculating seeding and soil amendment rates for the following year.

Humanitarian aid

Access to up-to-date imagery shows the creation of the Zaatari refugee camp over a nine-day period in July 2012. Designed to hold over 60,000 people, its population skyrocketed to over 150,000 before new camps relieved some of the pressure. The story map The Uprooted *tells the tale.*

Forestry

Dynamic access to data on forests in Europe is derived from the Corine Land Cover 2006 inventory. Corine means "coordination of information on the environment."

Mining

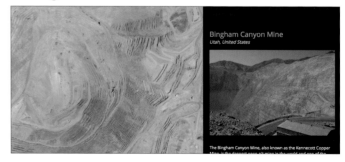

The geologic nature of the landscape comes to life using earth-orbiting satellites.

Natural disaster assessment

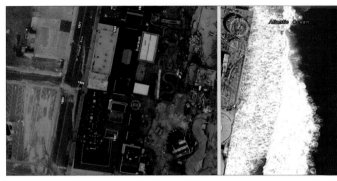

This scene shows the destruction of Hurricane Sandy's storm surge in Seaside, New Jersey. The active swipe map compares pre- and postevent imagery from the National Oceanic and Atmospheric Administration (NOAA).

Climate and weather study

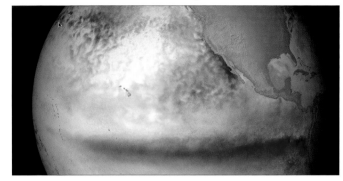

This short map presentation from NOAA answers many of the questions about the effects of El Niño. Scroll down to learn more about this climate feature and its characteristics.

Engineering and construction

Development projects actively under construction in the City of Pflugerville, Texas, are displayed here.

Oil and gas exploration

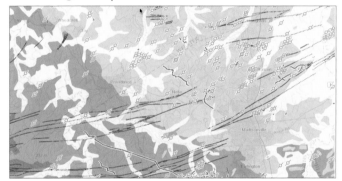

This geologic map compiled by the Kentucky Geological Survey relates themes of land use, environmental protection, and economic development.

Urban planning

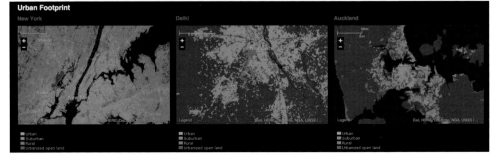

The Urban Observatory is an ambitious project led by TED founder Richard Saul Wurman to compile data that allows comparision of metro areas at common scales.

Measuring reflected solar energy
Passive sensors capture light reflected across the spectrum

A passive imaging sensor captures energy reflected or emitted from the scene it views. Reflected sunlight is the most common source of electromagnetic energy measured by passive sensors. These sensors provide the ability to obtain global observations of Earth and its atmosphere.

This map features MODIS satellite imagery from May 1, 2014, near the South Sandwich Islands in the southern Atlantic Ocean. The false color under the spyglass emphasizes snow and ice versus cloud cover.

A color infrared image, composed of near infrared, red, and green energy displayed as red, green, and blue highlights broad leaf and/or healthier vegetation as deep red hues, while lighter reds signify grasslands or sparsely vegetated areas.

Natural color is the workhorse of imagery. Well suited for broad-based analysis of both terrestrial and underwater features, urban studies, and reconnaissance, natural color imagery is the most familiar to a broad audience, and thus the most easily understood.

Higher-resolution panchromatic images are created when the imaging sensor is sensitive to a wide range of wavelengths of light, typically spanning the entire visible part of the spectrum stored and displayed as a single-band grayscale image. This enables creation of smaller pixels on the sensor, and a sharper image than the typical multispectral sensors on the same system.

Measuring transmitted energy
Active sensors send and receive their own signals

An active sensor is an instrument that emits energy and senses radiation that is reflected back from the earth's surface or another target. It is used for a variety of applications related to meteorology and atmosphere, such as radar to measure echoes from certain objects (such as rain clouds), lidar for capturing detailed surface elevation values, and sonar to measure seafloor depth.

Lidar

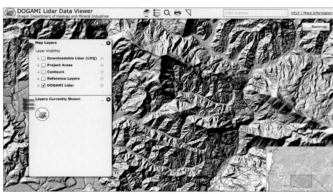

Lidar is collected from systems mounted on aircraft that gather up to 500,000 points per second, creating large collections of dense and accurate elevation points over a large area. The state of Oregon publishes free downloadable lidar, recognizing the community value.

Radar

Radar data has two primary strengths: it works in the dark and it can see through clouds. This makes it ideal for intelligence gathering and weather tracking, like this Next-Generation Radar (NEXRAD) application.

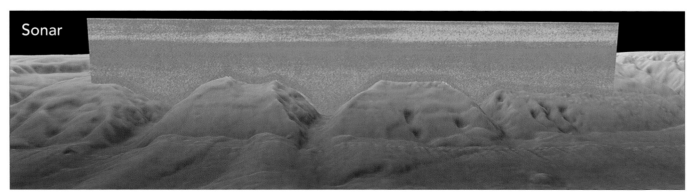

In a global initiative led by NOAA, water column sonar data is collected by active acoustic (or sonar) technology from the near surface to the seafloor. Since the ships are moving during the collection, the actual data ends up resembling a curtain or sheet as seen here. NOAA and the other contributing academic and international fleets are making the data available around the world to researchers and the public through this app.

Eyes in the sky
The constellation of earth-orbiting satellites

There are over 3,300 earth-observing satellites orbiting the globe, and the number is growing continuously. These myriad "eyes in the skies" are delivering an unprecedented payload of image data into the hands of spatial analysts, finding application to virtually all aspects of human activity. They cover low, medium, and high (geosynchronous) earth orbits. They're operated by government agencies (like NASA and the European Space Agency) and by private companies (like Digital Globe and Airbus). They cover all of the segments of the electromagnetic spectrum from ultraviolet to natural color to near, mid, and thermal infrared, and active microwave sensors such as radar.

But space is getting crowded. In addition to the 3,000-plus active spacecraft, the world's space agencies collectively track another 10,000-plus pieces of "space junk"—the spent boosters, battery-dead satellites, tools dropped by astronauts, and other debris from various events and mishaps.

As private launching and microsatellites gain favor, we can expect the number of sensors to continue to grow. The increasingly dense sensor grid offers promise for a wide range of applications, but it will bring serious challenges when it comes to effectively utilizing and disseminating the unprecedented flow of raw information.

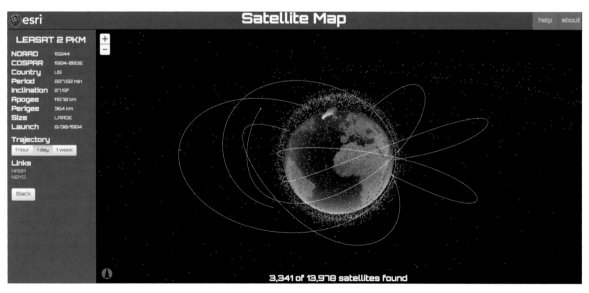

This application maps the current location of about 14,000 man-made objects orbiting Earth. The data is maintained by Space-Track.org, an organization that promotes space flight safety, protection of the space environment, and the peaceful use of space by sharing situational awareness information with US and international satellite operators and other entities.

Other perspectives
Obliques and street level imagery chart a new course

Not all geography is from the top down. Oblique-angle views provide a unique perspective that has particular application in reconnaissance and real estate, to name but two application areas. Street-level imagery, popularized by Google Street View, is another rich form of spatial data that creates an immersive and integrated navigation experience.

Obliques

Oblique aerial perspectives are able to depict the fronts and sides of buildings and locations on the ground. These perspectives can be stitched together to create composite aerial maps that seamlessly span many miles of terrain.

Street level

Mapillary is a platform that turns street photos into 3D maps for extracting geospatial data. Photos taken with mobile phones or consumer-grade cameras can be stitched together and reconstructed within minutes for upload to Mapillary and then ArcGIS.

Reality Lens

HERE Reality Lens is powered by high-quality, street-level panoramic views and high-precision lidar data that allows you to perform accurate measurements on street-level data.

Image resolution versus ground accuracy

An important concept in imagery is that of ground resolution. Every image has a ground resolution, typically expressed as distance on the ground. The imagery community refers to this as ground sample distance (GSD). This cell resolution is a measure of a square cell's height and width in ground units such as feet or meters.

A car is represented with three different pixel sizes or GSDs but displayed at the same scale. The more pixels on the car or the smaller the GSD, the better fidelity to resolve the car. On the left you can identify that it is a sedan; on the right you can merely detect that there is an object.

This image of the historic Mission Inn hotel in Riverside, California, is captured at approximately one-foot resolution. Click the image and zoom in as far as you can get. Each pixel represents about one foot on the ground. This type of imagery is appropriate for site-specific investigations and analysis.

This image of the same area is captured at one-meter resolution. The difference in resolution is significant. One-meter resolution data is appropriate for capturing and analyzing phenomena across larger areas of interest.

Thought Leader: Kass Green

Imagery reveals its character and structure in complex ways

Like a great piece of artwork, imagery reveals its character and structure in complex ways—always awe-inspiring, sometimes subtle, sometimes puzzling. First comes the astonishment of its raw beauty—stark glaciers in Greenland, the delicate branching of a redwood's lidar profile, a jagged edge of a fault line in radar, the vivid greens of the tropics, the determined lines of human impact, the rebirth of Mt. St. Helens' forests, the jiggly wiggly croplands of Asia and Africa, the lost snows of Kilimanjaro. Each image entices us to discover more, to look again and again.

After the first glimpse, we begin to explore. What's creating that unique spectral response? Why are the trees on north-facing slopes and shrubs on south-facing slopes in this area? Are the locations of different tree species related to slope and elevation? Why did this house burn and the one next door is untouched by flames? How many people live in this village? What crops are grown here? Will there be enough food to feed these people? How did the landscape change so dramatically? Who changed it?

Then, through the power of GIS, we discover the connections. If we're lucky, we travel to the field with our Collector apps to see for ourselves how the landscape varies in relationship to the imagery and other GIS layers. We use ArcGIS to organize and coregister the layers of information, and we mine for the variables that are most predictive. We learn how to tease out information about each object's location, height, shape, texture, context, shadow, tone, and color from the imagery and GIS data. And then we make maps—we inventory resources and monitor how they change over time.

Imagery has been my ticket to the world. Through it I have traveled the globe, heard amazing stories, and met fascinating people—all passionate about their endeavors and their communities. I am very fortunate to have found the beauty of imagery, and through it discovered the work I was clearly meant to do.

As founder of one of the first commercial companies to process Landsat data, Kass Green has been a leading voice for remote sensing and GIS for over 20 years. Her new Esri Press book Imagery and GIS: Best Practices for Extracting Information from Imagery *is scheduled for publication in 2017.*

Quickstart

The best way to get going is to first get a sense of how imagery can be leveraged in the ArcGIS platform by seeing it in action solving real problems (or at least informing those problems). The following story maps provide guided, curated views into the world of imagery and its important application to solving some of the planet's most pressing problems.

At the end of each story map, you'll find links to the source data that was used and some best practices for getting the data working properly in ArcGIS.

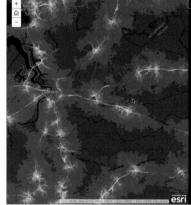

Learn ArcGIS Lesson

Study ecological land-use patterns in 3D using data from the Living Atlas in ArcGIS Earth

▶ Overview

The Global Ecological Land Units (ELU) map portrays a systematic division and classification of the biosphere using ecological and physiographic land surface features. Because it's a global dataset, it is an ideal data source to analyze using ArcGIS Earth.

In this lesson you'll open ArcGIS Earth, a lightweight app to access and display ELU data that will reveal patterns of change on Earth's surface. You'll analyze different areas of the planet, and see how well your own notions of these areas compare to the actual empirical data.

Build skills in these areas:
- Navigating ArcGIS Earth
- Loading data from the Living Atlas
- Accessing 3D KML data

▶ What you need:
- ArcGIS Earth
- Estimated time: 15 minutes

Start Lesson

Esri.com/imagerybook/Chapter2_Lesson

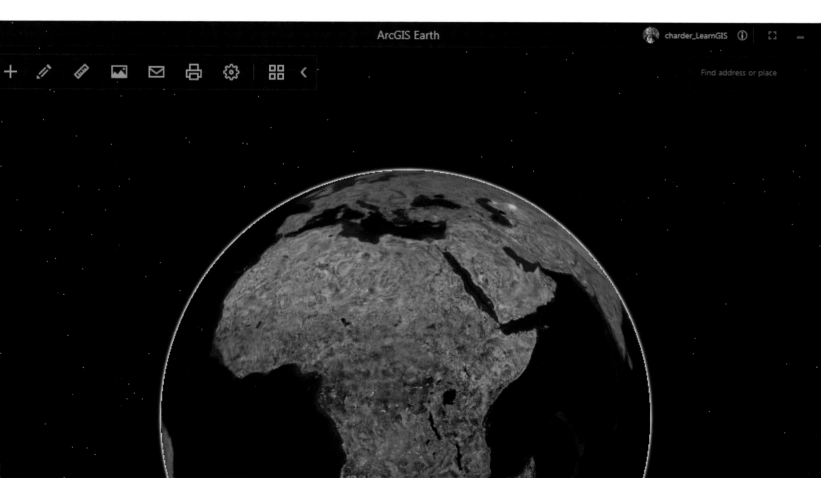

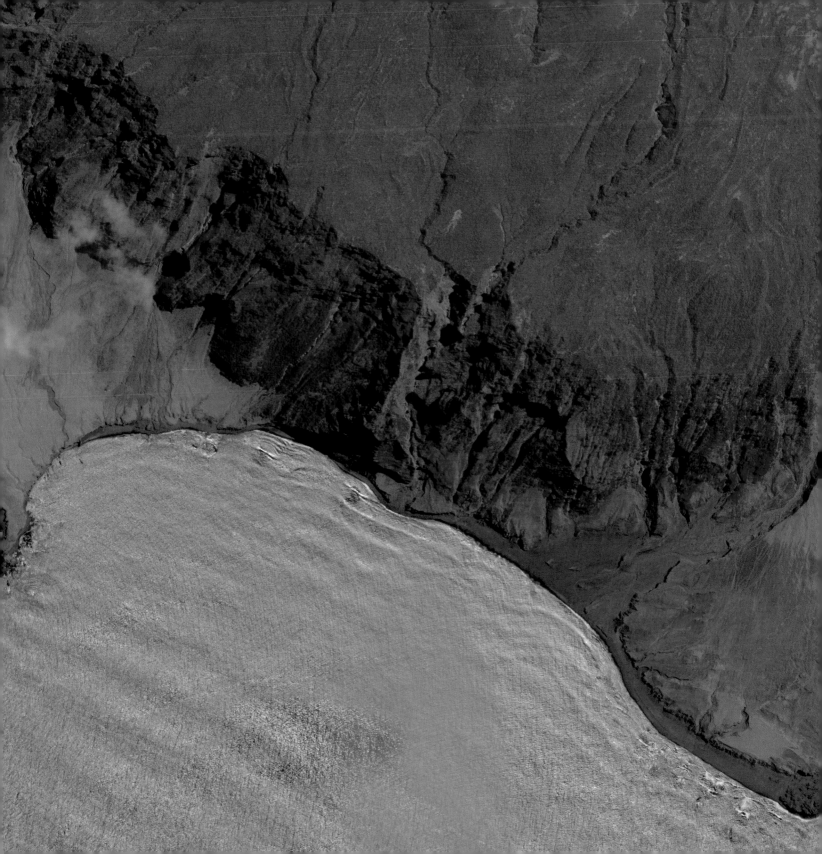

03

Defining Imagery

GIS and imagery are synergistic

For professional users, imagery and GIS have been perceived as closely related conceptually as complementary forms of digital geography, but still largely independent. Users had one system for GIS and another for image processing. An emerging big idea is that the two separate threads are now essentially interwoven within ArcGIS, resulting in a far-reaching and quite sudden expansion of image applications within the world of traditional vector-focused GIS.

A GIS cornerstone

Anatomy of an image

Imagery has long been a cornerstone for GIS that contributed synergistically to a wide range of GIS applications. In a very real sense, the broad and steady adoption of GIS over the decades has been fueled by imagery and remote sensing. Ideal for creating photographic basemaps and a perfect foundation on which to extract, trace, or otherwise digitize geographic features, imagery is the perfect complement to vector GIS, which used points, lines, and polygons to represent geography.

ArcGIS is a comprehensive image integration machine that opens the door to using the thousands of aircraft-, satellite-, drone-, and ground-based image sensors operating around the clock and around the globe. These digitally captured observations fit into geographic space and are time-stamped for temporal applications. The resulting information layers are being continuously added to the collective GIS knowledge of the planet, enabling people who work with geographic information to do more, and to do it faster and with wider impact.

In addition to its traditional GIS capabilities, ArcGIS also incorporates comprehensive image processing system capabilities that support the application, use, and integration of imagery and remote sensing.

This synthesis is founded on a series of key unifying concepts that link imagery to GIS. This chapter reviews these key concepts, reinforcing them with examples that help to communicate the power of imagery in your GIS.

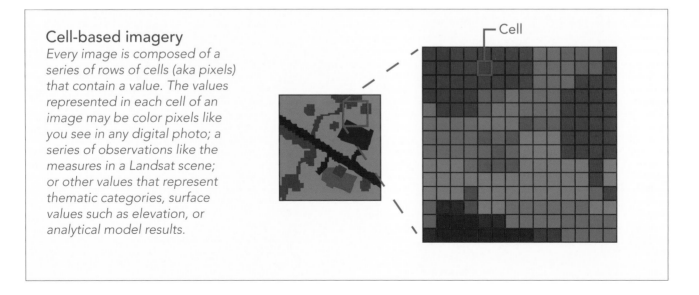

Cell-based imagery

Every image is composed of a series of rows of cells (aka pixels) that contain a value. The values represented in each cell of an image may be color pixels like you see in any digital photo; a series of observations like the measures in a Landsat scene; or other values that represent thematic categories, surface values such as elevation, or analytical model results.

Cell

Imagery layers are universal and varied

A raster is a grid of cells in a geographic space. The spaces within the grid are the cells. In a GIS, these cells are referenced to real geography. This cell-based raster structure is used to store and manage all imagery data. This fundamental grid structure makes raster data universal and useful for the representation of virtually any kind of geographic information. This means that all kinds of data can be integrated with imagery for mapping, advanced analysis, and data management.

True color aerial

Imagery along the Pelorus River in Marlborough, New Zealand. This is part of a rich national imagery dataset provided by Land Information New Zealand (LINZ).

Landsat scene

Landsat 8 scene of the mountains and canyons of Utah. This shortwave infrared image is useful for studying vegetation health, change detection, disturbed soils, and soil types.

Elevation surface

Elevation surface of Mount St. Helens derived from a satellite-acquired digital elevation model (DEM) and used to create a realistic hillshade.

Land cover

Land cover rasters identify different types of developed areas, agricultural lands, forests, and natural vegetation. Each cell represents the predominating value covering that cell.

Precipitation

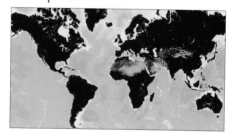

This time series map derived from MODIS satellite imagery contains a historical record showing the volume of precipitation that fell during each month from March 2000 to the present.

Flood inundation zones

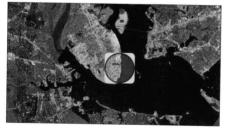

This map of Charleston, South Carolina, compares areas that are vulnerable to coastal flooding with the same areas in a US Coast Survey map of the city from 1863.

Under the hood

Imagery contains metadata

Earth observation imagery, like any digital photograph, contains important metadata that enables intelligent use of your information in ArcGIS. Software algorithms use this information to automate many of the once-cumbersome technical steps of georeferencing imagery.

Digital photo metadata

File Name:	IMG_0400.PNG
File Type:	Portable Network Graphics (*.png)
Location:	N44.95080° W93.36773° Datum: NAD83
Size:	2.58 MB
Created:	Oct 31, 2015 11:22:00 AM
Modified:	Oct 31, 2015 11:22:00 AM
Dimensions:	750 x 1334
Bit Depth:	24

Your digital photos contain metadata about the photo, including the date that the photo was taken, along with the location of the camera—its geotag, which records the GPS coordinates.

Imagery metadata

Similarly, drone, aerial, and satellite imagery contains detailed metadata items that enable more intelligent use—the spatial reference (or location) of your image, creation date, amount of cloud cover, and other properties.

ArcGIS puts this information to work with your imagery, creating automated, intelligent displays and analytics.

Images have a geographic reference

The defining characteristic of GIS data is that all layers are referenced onto the surface of the earth (or other planets, if that's your study area). Imagery data also has a spatial reference that enables it to be overlaid and used with all other GIS layer types. This is what makes ArcGIS a complete image integration platform.

Rasters have a spatial reference that enables them to be registered onto the earth's surface and to be combined with other GIS data layers.

Coordinate system

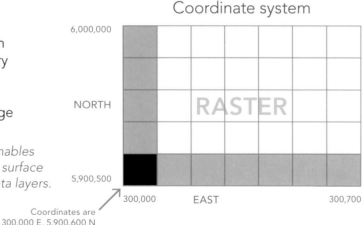

Coordinates are
300,000 E, 5,900,600 N

Geography is an organizing key
Imagery aligns with other geographic layers

Images are GIS layers, too. Like all geographic information, they are georeferenced to a location on Earth, which means they are registered with the other geographic layers in the GIS. This overlay capability is the fundamental concept upon which GIS operates. When combined with other mappable data, imagery transcends its status as merely a picture and becomes a true information source—data that can be combined, compared, and analyzed with any other data layers for the same area.

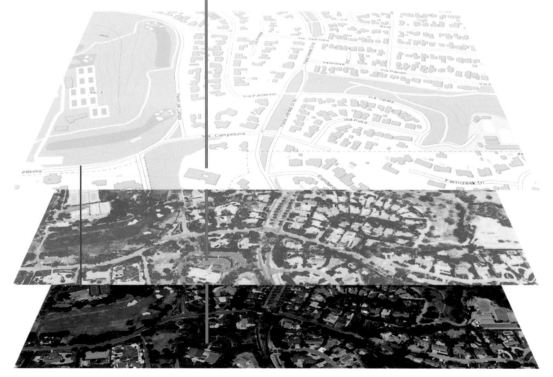

33.746851, -118.321296

All layers register and align in a GIS, including imagery layers. Georeferencing of information in this way is the hallmark of GIS and allows disparate types of information to be displayed, combined, and analyzed in a common geographic space.

An adaptable format
Any GIS layer can be represented as a raster

Once you realize rasters are just geographically aligned image files, they become the basis for a simple, universally applicable data format. All imagery is managed simply as collections of rasters. In similar ways, virtually any GIS dataset—vector features, continuous surfaces, and time series information—can also be represented using rasters.

In this way, GIS helps to organize and make sense of imagery. Unique datasets for a specific expanse of geography (its scale and extent) form a layer stack (sometimes called a data cube), which enables you to integrate an unlimited collection of independent layers. Consequently, imagery provides many of the layers in every GIS and adds extraordinary power.

Rasters can represent surfaces

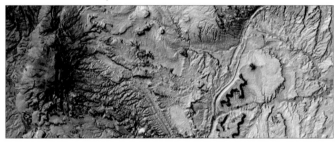

This example shows the extremely variable and dramatic elevation surfaces of southern Utah using the Landsat Shaded Basemap.

Rasters can represent features

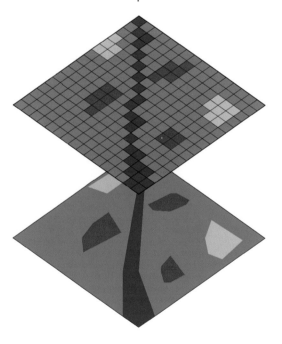

Rasters can represent time

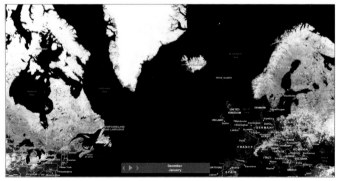

This web mapping application is designed to show the monthly average snowpack depth for the year 2014.

This diagram shows a stream segment as a vector (the variable-width blue polygon in the lower view) converted into a raster (along with other features) in the upper view.

Orthorectified imagery
Using elevation to enable accurate image georeferencing

Imagery has an amazing amount of information, but raw aerial or satellite imagery cannot be used in a GIS until it has been processed such that all pixels are in an accurate (x,y) position on the ground. Photogrammetry is a discipline, developed over many decades, for processing imagery to generate accurately georeferenced images, referred to as orthorectified images (or sometimes simply orthoimages). Orthorectified images have been processed to apply corrections for optical distortions from the sensor system, and apparent changes in the position of ground objects caused by the perspective of the sensor view angle and ground terrain.

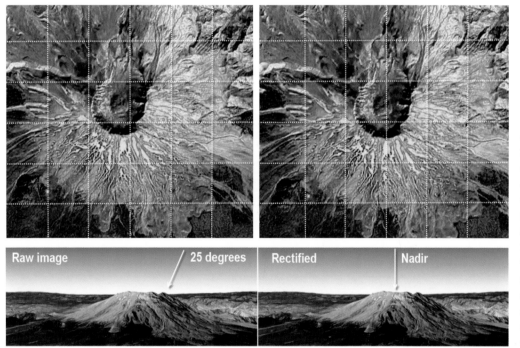

A view captured from an oblique angle (for example, 25°, left) must be corrected for relief displacement caused by terrain to generate the orthorectified view (looking straight down, right). Orthoimagery is produced by calculating the nadir view for every pixel.

The orthorectification process requires: An accurate description of the sensor, typically called the sensor model; detailed information about the sensor location and orientation for every image; and an accurate terrain model, such as the World Elevation service available from ArcGIS Online. After imagery has been orthorectified, it can be used within a GIS and accurately overlaid with other data layers.

Multispectral imagery
Enabling extrasensory perception

One of the most extraordinary types of imagery collected by remote sensing is multispectral imagery. Each image is composed of data from a series of onboard sensors that collect small slices (or bands) across the electromagnetic spectrum. The table below shows the complete list of wavelengths (expressed as bands) that are collected by the Landsat 8 imagery according to what they capture. The images below are examples of what you "see" by combining different bands into red, green, and blue electronic displays or hard-copy prints.

Band 1	Coastal Aerosol	Band 4	Red	Band 7	Shortwave Infrared 2	Band 10	Thermal Infrared
Band 2	Blue	Band 5	Near Infrared	Band 8	Panchromatic	Band 11	Thermal Infrared
Band 3	Green	Band 6	Shortwave Infrared 1	Band 9	Cirrus		

Image bands for Landsat 8 can be combined to create a number of scientific data layers used for research and analysis. For details, visit USGS Landsat online.

Natural color

The Natural Color (bands 4, 3, 2) combination of red, green, and blue is well suited for broad-based analysis of both terrestrial and underwater features and for urban studies.

Color infrared

Color infrared photography, often called false color photography because it renders the scene in colors other than those normally seen by the human eye, is widely used for interpretation of natural resources.

Land and water interface

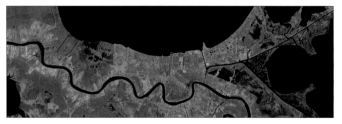

Landsat GLS Land and Water Boundary (bands 4, 5, 3) emphasizes the edges between land and water.

Vegetation analysis

This 6, 5, 4 band combination shows irrigated vegetation as bright green. Soils appear as tan, brown, and mauve.

Mosaic datasets
Collections of images

The recommended data structure within ArcGIS to manage and process imagery is the mosaic dataset. A mosaic structure enables significant big data capabilities for large, even massive, image collections. Each mosaic is composed of a number of related raster datasets, enabling you to keep your original individual image files on disk and to access them as part of a larger, integrated single collection. Mosaics are used to create a continuous image surface across large areas. For example, among other scenarios, you can use mosaics to handle coverage of very high-resolution image files for an entire continent. Or you can manage an entire historical map series for a nation for every year and every map scale. You can even manage huge multidimensional collections of time series information for earth observations and climate forecast modeling (often referred to as 4D). Creating mosaics is straightforward. You can point to a series of source georeferenced image files and automatically assemble a mosaic in minutes where each image acts as a tile within the collection.

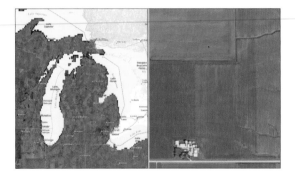

Managing extremely large collections
On the left is a mosaic overview for areas around Michigan, Illinois, and Wisconsin; a rural view of farms is on the right. This mosaic dataset by the National Agriculture Imagery Program (NAIP) contains well over 400,000 individual image tiles and covers the continental US. It includes the full information across multiple bands for each NAIP image as well as overviews for working with imagery at multiple scales.

Bringing historical collections to life
Image mosaics can also be made up of scanned historical maps, like the Historical Topographic Map Explorer containing 175,000 historic USGS maps accessible as an image mosaic in ArcGIS Online. You can also georeference your own historic maps and early aerial photographs and assemble them into mosaics.

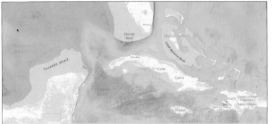

Analyzing multidimensional data
Multidimensional data is captured by location, by ocean depth, and by date. Earth scientists often refer to this data as 4D because it represents location in three dimensions with time as the extra dimension. Mosaic datasets help to manage and apply multidimensional data.

Rasters facilitate analysis
Assembling layer stacks

Rasters facilitate a broad array of sophisticated spatial operations and mathematical functions by providing a simple, universal data format that facilitates virtually any kind of geographic dataset. In turn, these enable simple workflows for performing all types of interesting and complex analytical operations and computations. When raster cells are piled on top of each other, they become a kind of processing-enabled data "stack."

Any GIS data layer can be turned into a gridded dataset, assembled with other datasets, thus creating a stack that can contain many layers, enabling you to combine data in useful analytic models.

Elevation
↓
Land cover
↓
Slope
↓
Soils
↓
Map algebra & analytic functions
↓
Your analysis model result

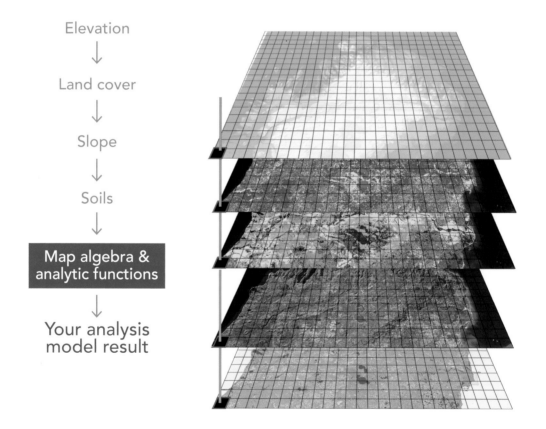

Rasters enable powerful analytic capabilities. For example, rasters stack on top of one another, enabling integration and useful overlay operations. And neighboring cells within a raster can be used for calculating zonal statistics, proximity to selected features, surface modeling, and flow functions. In addition, both 3D and time can be enabled analytically in significant ways.

Combining rasters in models
Stringing together a sequence of operations

In ArcGIS, raster layers and tools are combined into a progressive model. Each raster analysis tool performs a small, yet essential, operation on geographic data, such as combining layers with a weighted overlay, calculating the distance from each cell to specific features, or tracing flow paths across a surface. In turn, these derived layers can be fed into additional tools that generate further results. This enables you to string together a sequence of operations and create your own spatial analysis algorithms. With these you can use ArcGIS to model just about any kind of spatial problem you can think of.

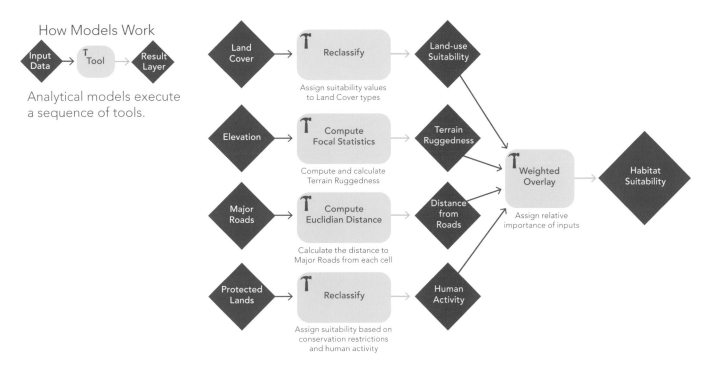

How Models Work

Analytical models execute a sequence of tools.

Geoprocessing is the methodical execution of a sequence of operations on geographic data to create new information. The raster data type has some of the richest tools for modeling and combining raster layers. The simplified habitat model for mountain lions above is an example of the remarkable modeling and analytical capabilities of rasters in ArcGIS.

Case study: A new level of cool
The Arctic Ocean Basemap

Over the past few years, Esri's Ocean Basemap team has noted the world's scientific attention shifting north. The receding sea ice and increased vessel traffic within the Arctic Ocean is coming front and center in discussions within the marine and maritime communities. To support the communities, Esri's Ocean Basemap team developed the Arctic Ocean Basemap.

The Arctic Ocean Basemap uses a polar projection that is optimized for this part of the planet. The Lambert Azimuthal Equal Area projection centered over Alaska allows the Arctic Ocean Basemap to interface seamlessly for polar-centric applications. It currently features data from numerous oceanographic sources of authoritative bathymetric data from Esri's close-knit maritime community. Like the Web Mercator version of the World Ocean Basemap, the Arctic Ocean Basemap consists of two map services. In this web map, the base and reference services are combined to create a map "sandwich."

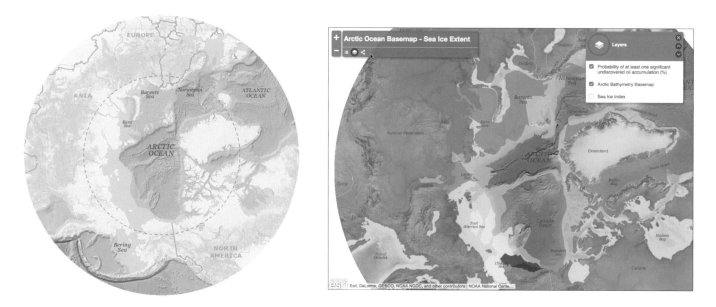

The Arctic Ocean Basemap (left) uses a special projection that is optimized for study of this region covering the northern latitudes of the globe from 90 to 50 degrees north. This (and the companion imagery version) is designed to be used as a basemap for overlaying other data for the Arctic region, such as this example (right) featuring sea ice extent and oil exploration data on top of the polar basemap.

Thought Leader: Dawn Wright
Imagery from beneath the waves

More than 1,500 people have climbed Mt. Everest, upwards of 300 have journeyed into space, and 12 have walked on the moon, but only three have descended and returned from the deepest part of the ocean. Like Odysseus, we must go to the sea in ships to explore our inner world. To collect its imagery, we rely on sound sensors aboard and remote sensors above and below. We then rely on GIS to turn this wide variety of data into information we can use: to map and restore habitats, design protected areas, manage deepwater fisheries, model tsunami run-up for evacuation, respond to oil spills, improve port navigation, and understand how storms erode the coast. We also use it to discover. Our "new" frontier is more than three billion years old, yet we have barely mapped a tenth of it—and this at the level of detail of a hiking map at a state park.

Still, the future of exploration of the deep, dark sea looks brighter as it is upon us. We're developing better sensors and analytics along with ways they can bring out the best in each other. For years, sensors on satellites and aircraft have been great at seeing features on the water surface but not under the waves. Airborne sensors rely on electromagnetic energy, and the deeper down you go (that surface-to-bottom concept known as the water column), the more such energy becomes distorted and dissipated. Sound waves, however, can be transmitted through water both farther and faster. So we've relied on waterborne acoustic sensors to help us visualize the water column. We use the sound signal's intensity (backscatter) to resolve the shapes of objects and the character of the ocean floor. Heavily sedimentary areas are

Esri Chief Scientist Dawn Wright aids in advancing the agenda for the environmental, conservation, climate, and ocean sciences aspects of Esri's work.

usually nonreflective, for example, while lava flows from recently erupted underwater volcanoes tend to be glassy and extremely reflective. Metal objects such as sunken ships and downed aircraft are also reflective.

With acoustic sensing voyages, we've gained imagery at the finest levels of detail, but such systematic surveys have been too few. So we need remote sensing via underwater videography and photography as well, along with GIS to translate different dimensionalities, resolutions, and accuracies of data into imagery that allows for our true understanding of inner space.

 Watch a video: Dawn Wright and Jack Dangermond discuss GIS and oceans

Quickstart

Mining the Landsat mother lode

Landsat 8 data is available for anyone connected to the Internet. Amazon Web Services (AWS) has made Landsat 8 data freely available so that anyone can use on-demand computing resources to perform analysis and create new products without needing to worry about the cost of storing Landsat data or the time required to download it. All Landsat 8 scenes from 2015 are available, along with a selection of cloud-free scenes from 2013 and 2014. All new Landsat 8 scenes are made available each day, often within hours of capture.

Esri President Jack Dangermond describes the impact of Landsat

Landsat sees the earth in a unique way. It takes images of every location in the world to reveal Earth's secrets, from volcanic activity to urban sprawl. Landsat sees a broad range across the electromagnetic spectrum, including what's invisible to the human eye. Landsat takes images of every location on Earth once every 16 days, so we can see how places change over time.

Esri is actively participating in this initiative. Amazon is hosting one petabyte of Landsat imagery from USGS on the Amazon Web Services cloud, making it accessible and usable for the GIS user community. Esri has gone one step further and created a set of publicly accessible web services updated on a daily basis. Each day, the latest Landsat 8 scenes are added and made directly accessible, along with the previous scenes. These services are multispectral and temporal, providing not only the latest pretty picture, but also the full information content from Landsat.

Esri takes the Landsat imagery hosted on Amazon's AWS cloud and makes it convenient and accessible to the ArcGIS user community. Shown above are some of the off-the-shelf, ready-to-use image services available.

Learn ArcGIS Lesson

Assess Burn Scars with Satellite Imagery

During the summer of 2015, wildfires ravaged Montana's Glacier National Park. When the blazes subsided, the Montana Department of Forestry and Resource Management measuerd the burn scars to quantify the damaged area. Burn scar measurements provide a baseline for forest regeneration and vegetation succession. However, on-the-ground measurements can be difficult and impractical. Instead, satellite imagery can form the basis of the measurements.

In this lesson, you'll assume the role of a geospatial scientist working with the Montana Department of Forestry to analyze the damage in Glacier National Park. You'll first compare Landsat 8 imagery from before and after the fires. Then, you'll change the band combination of the postfire imagery to emphasize burn scars and make a qualitative judgment. Afterward, you'll quantify your assessment by calculating a Normalized Burn Index (a ratio designed to highlight burned areas) from the imagery. Lastly, you'll create a feature class to represent the burn scar, calculate its acreage, and publish it to ArcGIS Online to share with the department.

▸ Build skills in these areas:

- Displaying different band combinations
- Creating a custom band combination
- Calculating a Normalized Burn Index
- Publishing layers to ArcGIS Online

▸ What you need:

- ArcGIS Pro
- Publisher or administrator role in an ArcGIS organization
- Estimated time: 2 hours

Start Lesson

Esri.com/imagerybook/Chapter3_Lesson

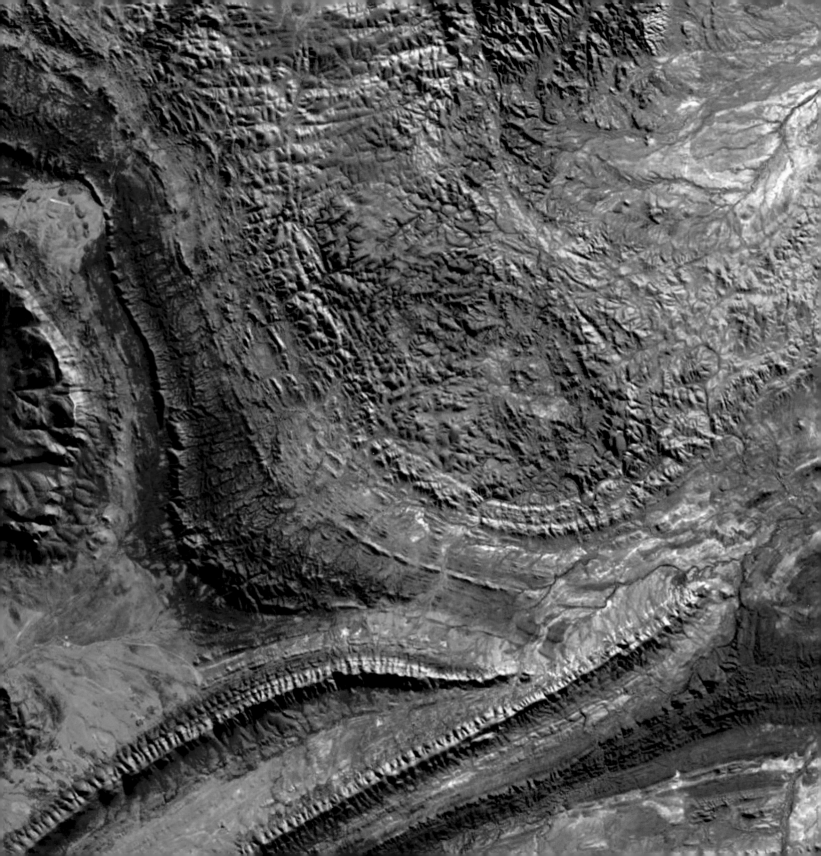

04

Perceiving the Imperceptible
Sensors give us superhuman eyes

Imagery provides more than plain pictures. Some sensors detect energy beyond what is humanly visible, allowing us to "see" across broad swaths of the electromagnetic spectrum. This enables scientists, geologists, farmers, botanists, and other specialists to examine conditions, events, and activities that would otherwise be hidden. The implications are profound and the applications are seemingly endless.

Expanding your point of view
Multispectral data increases your depth of field

Every day, the earth is directly imaged from scores of sensors in the sky and from orbit in space. Almost everything that happens is measured, monitored, photographed, and explored by thousands of imaging devices mounted on satellites, aircraft, drones, and robots. Much of this information ends up as imagery that is integrated into a large living, virtual GIS of the world, deployed on the web.

Some of these sensors see beyond what our eyes see, enabling us to view what's not apparent. Multispectral imagery measures and captures this information about a world that has many more dimensions than just the colors of the rainbow—it sees past the limits of what our eyes perceive.

Other active sensor technologies such as lasers and radar beam out signals that are reflected back at the speed of light, adding even more information to the collective repository. Some image sensors can see through clouds and under trees. Some detect things too subtle for any of our senses to distinguish. The richness and immediacy of this information is leading to a heightened understanding of the natural processes and human activities that influence our communities and environment. The ability to gather and exploit these new information sources is increasingly important to GIS practitioners.

This has been the mission of the remote sensing community since the first camera went up in an aircraft, and today the output from these sensors, across all the spectral ranges, is absolutely essential in enabling people to make better decisions.

The new kinds of multispectral sensors used in scientific work and analysis function the same way in GIS as the traditional natural light scenes; the basic principles are the same. These days, the speed and extent of collection and transmission means that the information is more immediate than ever, enabling us to make vital comparisons in near real-time after major man-made and natural events occur.

Web GIS is the nervous system for the planet, and imagery across the spectrum plays a vital role.

Nicaragua's Momotombo Volcano awoke with an explosive eruption in December 2015. This false color image highlights hot areas, primarily the lava flow that extends to the northeast.

Natural color imagery
The visible part of the spectrum

The camera on your phone is a sensor designed to capture photos—light and colors that represent objects we recognize in the way we are accustomed to seeing. These photos are collections of pixels expressed as the depth of red, green, and blue colors. Many aerial and satellite platforms capture images in this same way—along the visible spectrum—resulting in what are essentially georeferenced pictures of the earth from above. While not as exotic as some types of images, the value of natural color imagery is exceptionally high.

Simply capturing a series of pictures of the visible landscape from the air delivers fresh insight and helps us to understand many things within the framework of their geographic context and location. Additionally, the growing number of sensors and frequency of capture is increasing the application value of imagery and photography.

World imagery

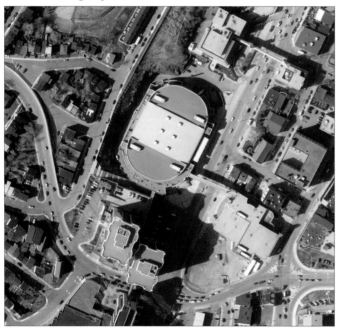

The World Imagery basemap (zoomed here to St. John's, Newfoundland) is the most commonly used basemap in ArcGIS, with over two billion requests per week. Stitched together from dozens of sources and stored in the cloud, this imagery is cached and optimized for use from global views down to high-resolution site views.

Patterns come alive

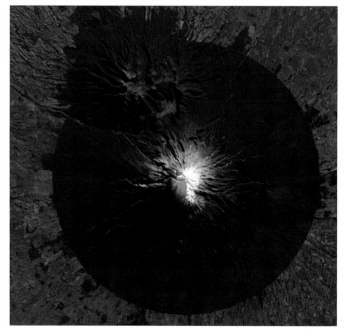

Egmont National Park, New Zealand, is home to Mount Taranaki, a dormant volcano. The nearly circular boundary, which protects forested areas on the mountain's slopes, is an example of patterns that emerge from exploring true color imagery. Click the image to explore a story map of circular patterns distributed across the globe.

The electromagnetic spectrum
Seeing beyond the visible

In the early history of powered aircraft, aerial photographs—pictures of the earth from above—began to be found useful for military and scientific applications. Quite quickly, imaging professionals and scientists realized that it was possible to detect beyond what is visible to the unassisted human eye. Deeper and richer information could be revealed by detecting waveforms from beyond the rainbow of visible light, into the invisible. As it turns out, these hard-to-detect realms of the spectrum offered some of the most meaningful insights. Hidden in these signals were previously unknown facts about Earth that have enabled us to understand our world far more effectively than had been possible.

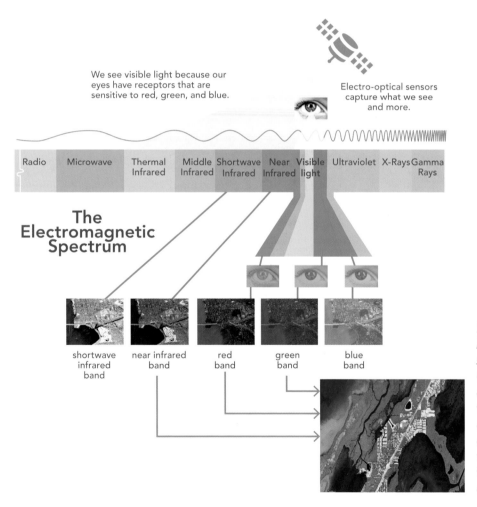

Many of these sensors measure bands across the electromagnetic spectrum and are known as Electro-Optic (EO) sensors. They record energy wavelengths from the sun that are reflected off or emitted from everything on the ground. These electromagnetic signals include visible light, infrared, and other frequency bands across the reflected energy spectrum.

Multispectral band combinations

Multispectral imagery measures different ranges of frequencies across the electromagnetic spectrum. One way to think of these different frequencies is as colors, where some colors are not directly visible to human eyes. These frequency ranges are called bands. Different image sensors measure different band combinations. The longest-running and perhaps most well-known multispectral imaging program has been Landsat, which began Earth image collection in the 1970s. By assigning data from three bands of the sensor to the red, green, and blue channels of an electronic display (or printer for a hard copy), color visualizations are created. Here are some examples of various alternate band combinations and their applications.

Panchromatic

Panchromatic imagery, commonly known as pan, is typically recorded at a higher resolution than the multispectral bands on any given satellite. It remains a critical source for many GIS applications as a reference for basic interpretation and analysis. Pan is often combined with other bands through a process called pansharpening to generate higher-resolution scenes.

Agriculture

In the Agricultural band (combination 5, 4, 1) vigorous vegetation appears bright green, healthy vegetation appears as a darker green, and stressed vegetation appears dull green.

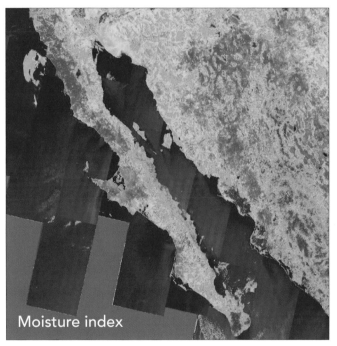

Moisture index

The Normalized Difference Moisture Index (NDMI) estimates moisture levels in vegetation where wetlands and vegetation with high moisture appear as blue growing to dark blue for higher moisture levels, and drier areas appear as yellow to brown shades. Image analysts often apply a formula to combine the selected multispectral bands to calculate various indexes.

Multispectral imagery in action
Putting remotely sensed image data to work

Marine mammal detection

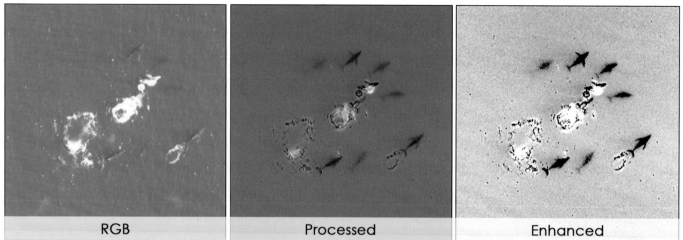

RGB   Processed   Enhanced

For those involved with marine mammal surveillance, infrared analysis in both daytime and nighttime conditions is an effective means of building accurate species inventories.

Coastal dynamics of erosion risk

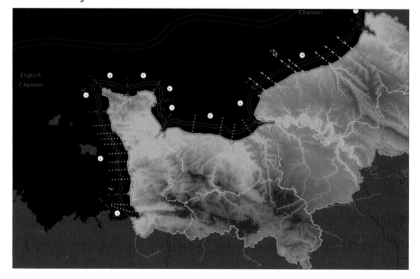

The French organization Réseau d'Observation du Littoral Normand et Picard uses imagery for several platforms to study the evolution of the coastline from Normandy to Picard. Presented in French, this stunning story map tracks the transit of sediments, sands, and gravel in the coastal strip as carried out by the action of tidal currents, waves, and prevailing wind. Coastal erosion is having a significant impact on the beaches and cliffs.

Monitoring severe floods

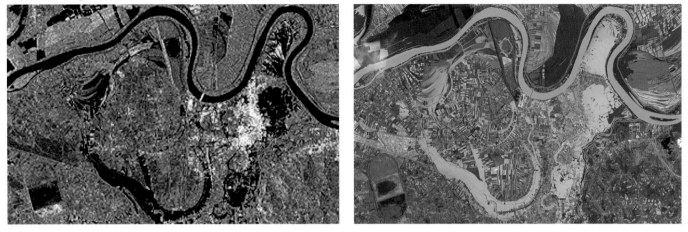

This comparison of two images in the west of Serbia shows two rivers overflowing extensively into the surrounding fields following a major flood in 2014. The towns of Krupanj and Obrenovac in Serbia are completely flooded. The ground and various parcels of land are completely hidden by water and mud. The image on the left is a TerraSAR-X radar image taken the night the flooding started, already showing breaches in the earthen dams, followed by an optical SPOT image (right) taken once the cloud cover reduced, and displaying the full devastation of the flood breach.

Mining imagery for mineral patterns

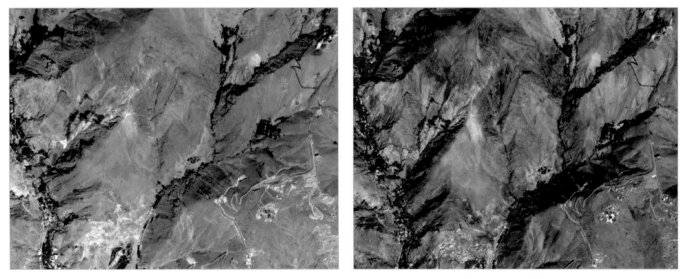

These views from near Tehran, Iran, show a natural color band image on the left and short-wave infrared (SWIR) image on the right. Note how one particular rock top pops out in pink using the SWIR bands and is not as easily discernible in the natural color band combination. The variation in rock types allows analysts to easily identify specific mineral patterns, greatly narrowing the search areas for particular materials.

Copernicus—Europe's eyes on Earth

Copernicus is the European Space Agency's (ESA) earth observation program to monitor how our planet and its environment are changing. Finite natural resources are under pressure from our global population growth, generating an ever-increasing demand for safe living space, fresh water, fertile land, and clean air.

To make effective decisions, public authorities, policy makers, businesses, and citizens need reliable and up-to-date information services. The Copernicus program is founded on a dedicated constellation of satellites named the Sentinels—more than a dozen will be launched into orbit over the next 10 years, covering marine, land, climate, emergency, security, and atmospheric applications. The first one—Sentinel-1A—is a polar-orbiting, all-weather, day-and-night radar imaging mission for land and ocean services. It came online in late 2014. The second radar satellite, Sentinel-1B, launched successfully in April 2016. Sentinel-2A was launched in June 2015 to monitor land and vegetation, as well as coastal waters. This new satellite carries a high-resolution optical instrument that covers 13 spectral bands with a swath width of 290 kilometers. The refined band sensitivity of the sensor means it's particularly adept at monitoring urban sprawl and land use, as seen in this image (right) of Rome, Italy, taken in August 2016.

You can download the ESA app to watch the Sentinel satellites orbit the earth live. Search "ESA Sentinel" in the iTunes App Store or Google Play on Android.

The combined image of the optical sensor, Sentinel-2A and the radar sensor, Sentinel-1A, reveals the urban development and potential land changes in Rome in 2016.

Sentinel-1A over Greenland

Copernicus program online

 Watch a video of the flight path of Sentinel-1A

Case study: The Landsat program

The Landsat program is the longest-running enterprise for acquisition of satellite imagery of Earth. On July 23, 1972, the first Earth Resources Technology Satellite was launched. This was eventually renamed Landsat. The most recent, Landsat 8, was launched in 2013. The instruments on the Landsat satellites have acquired millions of images. Archived at Landsat receiving stations around the world, these images are a unique resource for global change research, agriculture, cartography, geology, forestry, regional planning, surveillance, and education. Historical archives can be viewed through the USGS EarthExplorer website. Through Landsat 7, the data has eight spectral bands with spatial resolutions ranging from 15 to 60 meters. Every part of the Landsat coverage is rephotographed every 16 days.

Landsat 8 added two additional bands. The three key mission and science objectives for the latest "bird" were to collect and archive medium-resolution (30-meter per pixel) multispectral image data affording seasonal coverage of the global landmasses for a period of no less than five years; ensure that Landsat 8 data is sufficiently consistent with data from the earlier Landsat missions in terms of coverage and spectral characteristics, output product quality, and data availability to permit studies of land cover and land-use change over time; and distribute Landsat 8 data products to the general public on a nondiscriminatory basis at no cost to the user.

 Video: Landsat data continuity mission overview

In 1976, Landsat 1 (left, during final assembly) discovered a tiny uninhabited island 20 kilometers off the eastern coast of Canada. This island was thereafter designated Landsat Island after the satellite.

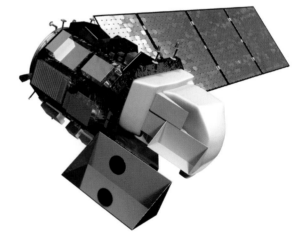

Landsat 8 delivered two new spectral bands, a deep blue coastal/aerosol band and a shortwave-infrared cirrus band, allowing scientists to measure water quality and improve detection of high, thin clouds.

The Baltic Sea faces many serious challenges, including toxic pollutants; deepwater oxygen deficiencies; and toxic blooms of cyanobacteria affecting the ecosystem, aquaculture, and tourism. The images from Sentinel-2 show features down to 10 meters across, revealing exquisite detail of an algal eddy. A ship (at point of arrow) can even be seen near the center of the eye. The ship's track is visible as straight dark feature where the algae have been disturbed by turbulence created by the ship's propellers as it mixes water in its wake.

Finding the so-called "red edge"
Drone-borne multispectral cameras are revolutionizing agriculture

Multispectral remote sensing serves up radically new perspectives on crop health and vigor. The red edge is the boundary between the (visible) red and (invisible to humans) near infrared (NIR), and it's called an edge because the spectral profile of vegetation shows a dramatic rise in brightness from red to NIR. When vegetation is stressed and the profile changes, the edge moves, and a narrow spectral band at the right wavelength can detect a dramatic difference. MicaSense's RedEdge cameras are calibrated for classic spectral bands of blue, green, red, and NIR, but also have a fifth band at 720 nanometers explicitly to detect movement of the red edge. An inexpensive drone equipped with a MicaSense camera can be flown as often as necessary, allowing the grower to finely calibrate irrigation, fertilization, and the application of insecticides (which can also be applied via drone).

RGB (natural color)

Color infrared

Normalized Difference Red Edge Index

NDRE

0.250
0.268
0.286
0.304
0.322
0.340
0.359
0.377
0.395
0.413
0.431
0.450

Advanced, lightweight, multispectral cameras provide accurate multiband data for agricultural remote sensing applications. What this means is that a hyperlocal drone flight can easily identify the stressed areas of a crop—the red area in the upper rows of the vineyard seen through the red edge filter (right).

Beyond reflected sunlight

Lidar, radar, and sonar—the active sensors

Space-based image sensors typically measure solar light reflected from the ground. This is often called passive sensing. In contrast, active sensors such as lidar, radar, and sonar emit pulses of energy and then monitor the return of energy. As the return energy arrives at the sensor, the intensity and time stamps of the return signals are used to determine the precise shape and location of the object. Active sensors work perfectly well at night, an inherent capability of active sensing technologies.

Lidar uses laser technology to scan objects and landscapes, and records the elevation surface detail and shape of these features based on their measured distance from the scanning device. Laser returns are used to generate a cloud of points in x,y,z space containing a series of attributes, such as intensity, look angle, and a very accurate time stamp. Lidar can be used to model not only the terrain, but also the tree canopy, buildings, power lines, bridges, and everything else on and above the surface.

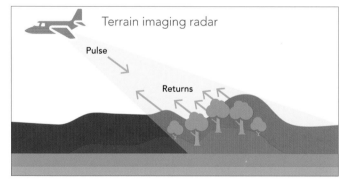

Radar is a particularly useful sensor when flying at night or in cloud cover. Unlike optical sensors that need clear, unobstructed views, radar works equally well at night and in inclement weather. The disadvantage of radar is that its resolution is limited by the radio wavelength.

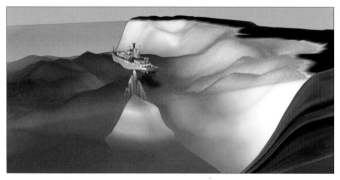

Sonar is the wave of choice in bathymetry. An acoustic pulse is emitted from a transducer and propagated in a single, narrow cone of energy directed downward toward the seafloor (or a lake bed or river bottom). A transducer then "listens" for the reflected energy from the underwater terrain, providing time returns that can be converted to depth measurements.

Lidar applications

High-resolution surface elevation mapping

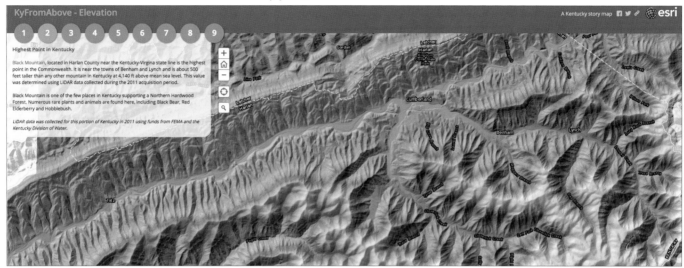

KyFromAbove is a statewide mapping program for the Kentucky state government that includes the comprehensive lidar collection of surface elevation at high resolution throughout the commonwealth.
This story map tells the tale of the statewide collection and how it is being put to use.

Digital surface model (DSM)

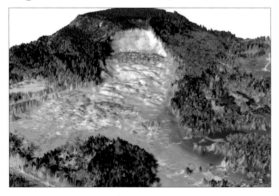

Lidar is often collected after major terrestrial events such as landslides. This 3D scene shows the impact of the massive and deadly Oso mudslide in Washington State, _news of which reached the world in the spring of 2013_.

Photorealistic scenes

LAS files (the generic lidar exchange format) are a collection of points, each with horizontal coordinates and a vertical elevation value. LAS files provide a common format for storing additional information such as laser intensity, scan angle, and return information. When encoded as red-green-blue (RGB), the scenes take on a photorealistic appearance, like this visualization from Petaluma, California.

On the radar
Reflecting at high frequencies

Weather surveillance radar (WSR) and Doppler weather radar

Weather-based radar represents a type of radar used to sense precipitation and type (such as rain, snow, or hail), as well as to calculate the motion of storm systems. Modern weather radars are mostly Doppler radars, capable of detecting the motion and location of rain droplets in addition to the intensity of precipitation. Radar data can be analyzed to determine the structure of storms and their potential to cause severe weather.

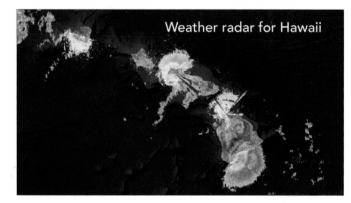

NEXRAD is a network of 160 high-resolution Doppler weather radar stations operated by the National Weather Service (NWS). This interactive map of NEXRAD radar enables you to view and interact with up-to-the-minute weather for the Hawaiian Islands.

SRTM elevation data

The Shuttle Radar Topography Mission (SRTM) was a NASA Space Shuttle-based research effort that obtained digital elevation models on a near-global scale from 56° S to 60° N in an effort to generate a complete high-resolution digital topographic database of Earth from space.

To acquire elevation data, the Space Shuttle *Endeavour* was outfitted with two radar antennas, one in the shuttle's payload bay and the other tethered on the end of a 60-meter mast. The radar instruments on board applied synthetic aperture radar, which was used to generate terrain surface maps of the earth at a resolution of 30 meters. Once the mission was complete and the data could be processed, it was shared publicly in the first few years at a reduced resolution of 90 meters. Recently, elevation data has been released for the world at the full resolution of 30 meters.

This web scene tour of Africa uses elevation data collected by the Shuttle Radar Topography Mission, which provided the first-ever global elevation coverage for the world.

Where's the heat? Where's the gas?
Thermal and gas sensing applications

All objects on Earth emit or radiate infrared radiation because they have a temperature. This energy is long wavelength and can be collected by thermal infrared (TIR) sensors. Thermal imaging is a day-night capable sensor since it does not require illumination; all objects radiate energy on their own, day or night. Objects that are hotter radiate more energy, so on a thermal image, they appear brighter.

Geothermal energy

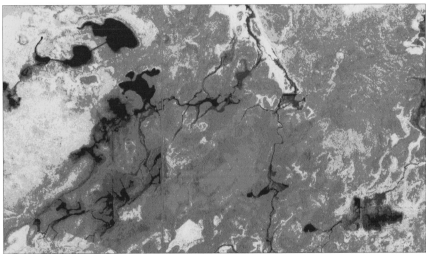

Thermal infrared data collected as part of a project funded by the US Department of Energy Geothermal Technologies Program. This particularly active geothermal region is in the Connley Hills of central Oregon.

Imagery collected using a camera with a thermal infrared sensor shows a toxic gas plume, which was invisible to the naked eye. This release from the mountainside in Porter Ranch, California, continued for 110 days and was finally plugged in February 2016.

Hyperspectral imagery
Fingerprinting the ground

Hyperspectral sensors see the world using a broad swath of the electromagnetic spectrum, but unlike multispectral sensors, the hyperspectral systems provide many more spectral bands, enabling observation of detailed spectral signatures. Hyperspectral images can enable identification of specific plants and minerals.

Many projects that use hyperspectral sensors are designed for specialized focus on particular bands to discover the presence of specific phenomena. These signatures enable identification of the materials that make up a scanned object. Detection of known spectral objects is aided by their tendency to have very similar spectral characteristics wherever they occur. For example, the spectral signature of a white pine tree is consistent and distinct from the signature of a sugar maple. Rocks that hold significant amounts of one mineral are distinct from similar-looking rocks holding another type of mineral. These distinctions are used to identify and extract features for use in a variety of applications.

Mineral mapping

Hyperspectral map of Cuprite, Nevada, provides a synoptic view of the surface mineralogy, and identified a previously unrecognized early steam-heated hydrothermal event that resulted in extensive distribution of iron-bearing elements.

Mapping sands and ocean substrates

Unique spectral signatures for different types of soil and sand can facilitate mapping for geology or planning mining prospects.

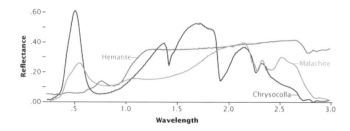

Individual materials scanned using hyperspectral imagery have unique characteristics, or fingerprints. This graph compares the reflectance of hematite (an iron ore) with malachite and chrysocolla (copper-rich minerals) from 200 to 3,000 nanometers in wavelength.

Thought Leader: Sarah Parcak

Using satellite archaeology to protect ancient sites

There may be hundreds of thousands, if not millions, of undiscovered ancient sites across the globe, and Sarah Parcak wants to locate them. As a satellite archaeologist, she analyzes infrared imagery collected from far above the earth's surface and identify subtle changes that signal a man-made presence hidden from view. Doing so, she and her colleagues aim to make invisible history visible once again—and to offer a new understanding of the past.

Inspiration comes from her grandfather, an early pioneer of aerial photography. While studying Egyptology in college, Parcak took a class on remote sensing and went on to develop a technique for processing satellite data to see sites of archaeological significance in Egypt. The method allows for the discovery of new sites in a rapid and cost-effective way.

In partnership with her husband, Greg Mumford, they have directed survey and excavation projects in various places in Egypt. She's used several types of satellite imagery to look for water sources and archaeological sites.

Her latest work focuses on the looting of ancient sites. By satellite-mapping Egypt and comparing sites over time, the team noted a 1,000 percent increase in looting since 2009 at major ancient sites. It's likely that millions of dollars' worth of ancient artifacts are stolen each year. The hope is that, through mapping, unknown sites can be protected to preserve our rich, vibrant history.

Sarah Parcak is a leading expert on space archaeology. She is from Bangor, Maine, and is a National Geographic Society Archaeology Fellow, Fellow of the Society of Antiquaries, and a 2013 TED Senior Fellow. Sarah serves as the founding director of the Laboratory for Global Observation at the University of Alabama at Birmingham, where she is a professor.

 Watch Sarah Parcak's TED talk

Bringing the map to the image
Working in image space

Not all imagery applications require the projection of sensor data onto a map or, in other words, the registration of imagery into a geographic coordinate system. There are many applications where it is more effective and appropriate to work with the original image and view it from the perspective of the camera. This is referred to as working in image space, in contrast to working in a map coordinate system. Numerous military and civilian reconnaissance applications involve the use of both a map view and an image window. For example, inspection applications effectively use an image view and a map view in concert.

Building inspection is one of the early useful applications to come from the drone revolution. In this example using drone imagery (flown, incidentally, with a 3DR sub-$1,000-priced model), the flight path is seen as a yellow circle and the blue dots signify the capture points. The oblique view in the Image Viewer is an example of viewing a series of images in image space to see undistorted views captured by the camera.

Full motion video
Put your video on the map and your map on the video

ArcGIS has the ability to integrate and incorporate full motion video (referred to as FMV), presuming you have metadata that describes the geographic location for your video. This is akin to how aerial imagery is georeferenced except that every frame in the video is georeferenced. Such georeferenced videos adhere to formats established by the Motion Imagery Standards Board (MISB), which oversees standards for full motion video capture pertaining to the defense and intelligence communities in the United States.

This enables MISB-compliant video frame locations to be placed as windows into your map views—and your map data as optional overlays in your video. FMV technology enables you to quickly and easily analyze video data from many kinds of airborne sensors—such as aircraft, drones, and other UAVs.

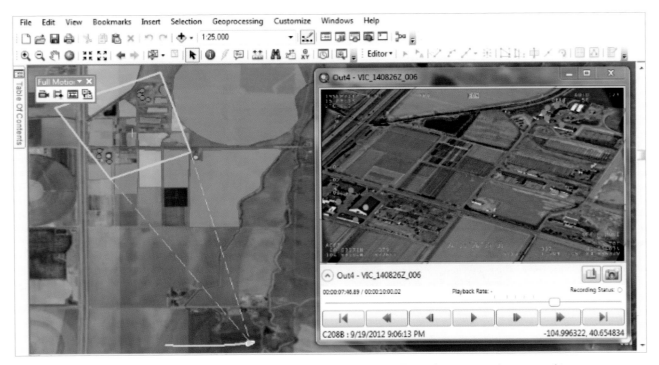

Full Motion Video capabilities link the locations and look angles of the drone (or other aircraft) camera to a GIS map, which allows you to track the location of the aircraft, and also the frame-by-frame footprint of the area being seen in the camera image.

Quickstart

Multispectral Imagery Gallery in the Living Atlas of the World

The quickest way to access the multispectral imagery in ArcGIS Online is through the Living Atlas of the World. But this is only a starting point. Once you've opened any of these services in ArcGIS, you can use the Display Image menu selection to alter the bandwidths and create your own combinations.

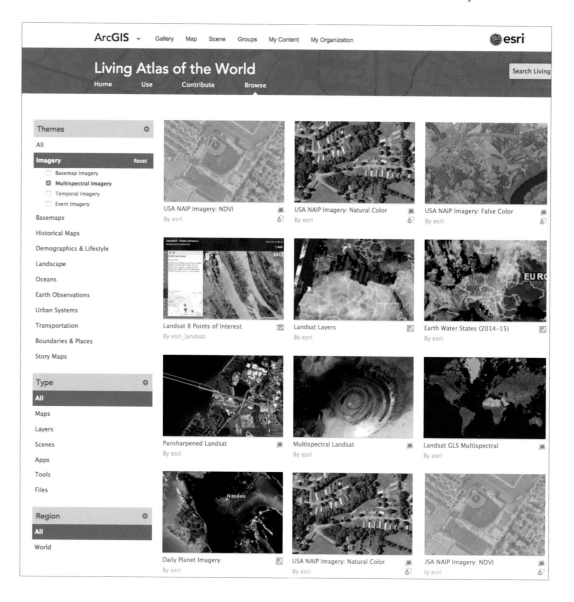

Learn ArcGIS Lesson

Water level analysis of Poyang Lake, China, using multispectral imagery to identify and measure water and land surfaces over a period of time

In these lessons, you'll assume the role of a geospatial scientist tasked with calculating the change in area of the lake between 1984 and 2014. Using Landsat imagery, you'll classify land cover in three images of the lake taken at various times over the past 30 years to show only the surface area of the lake. You'll then determine the change in lake area over time.

▸ **Classify Land Cover to Measure Shrinking Lakes**

Poyang Lake, China's largest freshwater lake, has always had significant seasonal fluctuations in water level. Fed both by rains and the Yangtze River, Poyang Lake has lately experienced even more extreme fluctuations due to several years of drought and the construction of the Three Gorges Dam.

Dry season water levels are alarmingly low, and even rainy season water levels have fallen. The changes have impacted the local economy and altered the land cover of the area. But if locals want to do something about their lake's disappearance, they'll need to back their live observations with scientific facts.

Three Gorges Dam in China.

▸ **Build skills in these areas:**
- Classifying land cover
- Calculating change in area

▸ **What you need:**
- ArcGIS Pro
- ArcGIS Pro Spatial Analyst extension
- Estimated time: 1 hour 15 minutes

Start Lesson

Esri.com/imagerybook/Chapter4_Lesson

Man walks in dry riverbed near Poyang Lake, China.

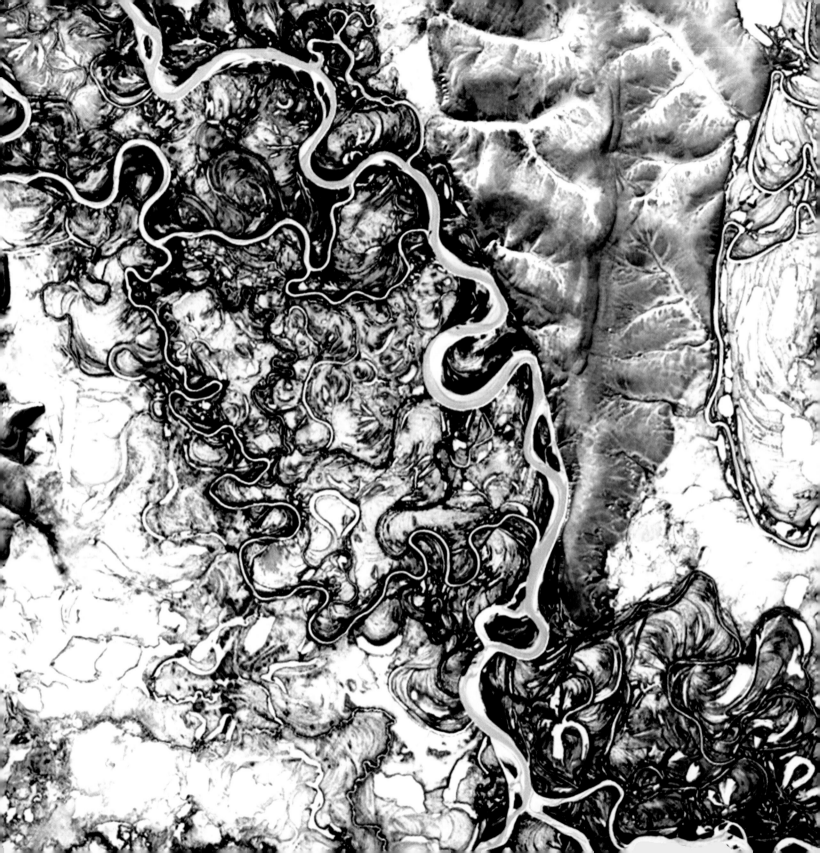

Turning Imagery into Information
Analyzing imagery to create understanding

Image analysis allows us to derive new understanding from existing data by creating analytic maps for insight and knowledge. These raster (cell-based) layers can be used to map and model virtually anything that happens across the earth's surface, like agriculture, planning, hydrology, climate, wildlife habitats, and much more. The big idea in this chapter is that imagery data—with its cell-based data stucture—allows for the systematic and controlled analysis of a vast array of phenomena across multiple layers.

Imagery analysis creates understanding
GIS with imagery opens the door to solving complex problems

ArcGIS provides an analytic platform that enables you to combine imagery with other kinds of geographic information within analytic models. It's simple. GIS organizes information as geographic layers. Meanwhile, earth imagery scenes and sensor data are also accessible as layers. ArcGIS provides thousands of analytic operators that can derive statistical information, model movement and flow across your surfaces, help you to combine layers to find the most and least suitable areas for your activities, and much more.

Imagery provides a versatile information feed—a virtual fire hose of information to your GIS. In turn, ArcGIS has a number of spatial analysis operators that enable you to gain deeper insight into and understanding of your information. These analytic tools enable you to address virtually any kind of question, such as deriving the statistical signal from your data, examining a sequence of events through time, and forecasting and predicting them into the future. Spatial analysis entails identifying and deriving new information layers to help solve all kinds of problems, such as finding the right places to build, analyzing your business performance or where that new market may be hiding in plain sight, evaluating and managing your agricultural production, or monitoring and forecasting diseases.

Virtually any problem we face can gain from analytical insight provided by ArcGIS. And imagery is always a critical information source in your analytic work.

GIS and imagery analysis have come together and integrated only recently. And with the advent of cloud and enterprise server computing, modern computing systems are capable of analyzing massive volumes of image information. Limits in modeling have been significantly reduced, enabling you to model and analyze your information in deeper, more profound ways.

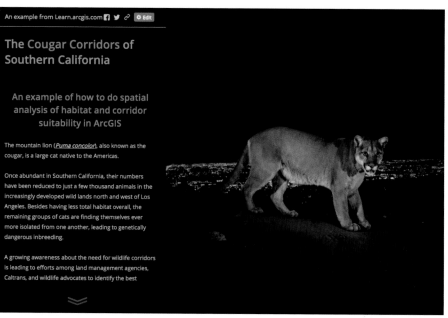

An example from Learn.arcgis.com

The Cougar Corridors of Southern California

An example of how to do spatial analysis of habitat and corridor suitability in ArcGIS

The mountain lion (*Puma concolor*), also known as the cougar, is a large cat native to the Americas.

Once abundant in Southern California, their numbers have been reduced to just a few thousand animals in the increasingly developed wild lands north and west of Los Angeles. Besides having less total habitat overall, the remaining groups of cats are finding themselves ever more isolated from one another, leading to genetically dangerous inbreeding.

A growing awareness about the need for wildlife corridors is leading to efforts among land management agencies, Caltrans, and wildlife advocates to identify the best

Suitability habitat modeling is a classic GIS and image analysis problem as described in this story map. Source data from multiple sensors is combined and georeferenced in a way that allows land-use planners to identify strategies such as wildlife corridors to protect the long-term survival of the species.

Traditional image analysis

The use of imagery for GIS analysis is nothing new. Throughout the past few decades, imagery sources such as multispectral layers, digital elevation models, and digital orthophotos have provided an analytical foundation for modeling and feature extraction. Here are some common examples.

Assessing current tree canopy

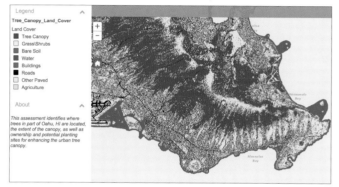

This analysis identifies tree cover and the extent of tree canopy across the island of Oahu, Hawaii. The USGS applied image analysis on Landsat and other data sources to derive land cover for the entire island.

Photogrammetry

Orthophotos are used here to map a port facility in Germany. Image interpreters capture accurate features from these types of imagery sources.

Land cover classification

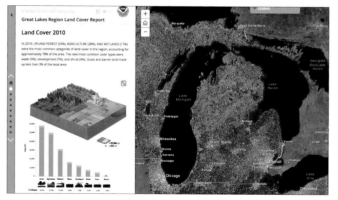

Changes in land cover and land use can tell dramatic stories about the rapid environmental changes that are occurring in places like the Great Lakes region of the United States.

Assessing crop health

Multispectral imagery can provide a new perspective on crop health and vigor. The Normalized Difference Vegetation Index (NDVI) reveals healthy potato and canola crops in Saskatchewan.

Navigation, flow, and surface modeling
Prebuilt for advanced analytics and visualization

Calculate cost surfaces

A cost surface is a raster grid in which each cell value represents the cost to travel through it. Cost surfaces can model things like the optimal path for a bushwhacking fire crew, predicting how a fire might spread, or predicting the travel preferences for how a mountain lion might move across its habitat range. In this map, green areas represent lower travel costs for the big cats in semirural Southern California.

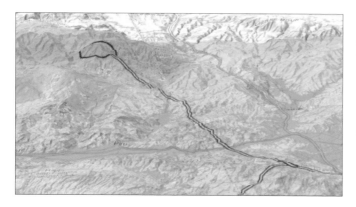

Find best paths

A cost path calculates the least-cost travel path for traveling from one location to another. Costs can represent a number of criteria, including actual monetary expenditure, but more often are related to time and effort required to complete the journey. In this example, you can see the best path for cougars to travel between two of their core habitat areas.

Visual and visibility analysis

A surface displayed in 3D space has value as a visual display backdrop on which to drape data and analyze it. This perspective scene shows a restored watershed and river draped on a digital elevation model of the terrain.

Characterizing the world's ecology
The ecological land units project

The very nature of cell-based data makes it ideal for certain kinds of advanced analytics that can't even be considered with vector data. The ecological land units (ELUs) project is one such example. Four global layers (bioclimate, landforms, rock type, and land cover) were overlaid and combined to create a single output surface that portrays a systematic division and classification of the global biosphere using ecological and physiographic land surface features to describe and characterize each land unit. "This map provides, for the first time, a web-based, GIS-ready, global ecophysiographic data product for land managers, scientists, conservationists, planners, and the public to use for global- and regional-scale landscape analysis and accounting," said Roger Sayre from the USGS.

Ecological land units are areas of distinct bioclimate, landform, lithology, and land cover that form the basic components of terrestrial ecosystem structures. The ELU map was produced by combining the values in four 250-meter cell-sized global rasters using ArcGIS. These four components resulted in over 3,900 unique combinations of ELUs worldwide. The ELUs and their four input layers represent the most accurate, current, globally comprehensive data available.

Describing or characterizing a place
Segmentation and classification

Imagery can be used to automate the classification and locations of land into specific categories, such as different types of land uses and land cover. These derived layers can then be used as basemaps and, more interestingly, in subsequent analyses. Classifying a series of images from different time periods also enables analysts to explore how a location is changing, whether from natural forces or human interventions.

Land cover change detection

This forest change analysis tool evaluates the total tree cover loss and number of active fires within the selected area of interest, and shows the results according to the various land cover classes. The Global Forest Watch change analysis tool uses spatial and temporal information to allow you to conduct your own investigation on forest cover change, current land cover, and legal classifications in your area of interest.

Segmentation

Image segmentation is defined as a process of partitioning an image into homogenous groups such that each region is homogenous. This map shows the impervious surface of each parcel after these surfaces have been segmented out using feature extraction analytics in ArcGIS. This is a classic segmentation application.

Suitability analysis
Finding the best locations

A common question that GIS analysis helps to solve is, where is the best place to put something? Suitability models are used for just that purpose—to find the ideal place to build or preserve, depending on the objective. The problems addressed can be wide-ranging: where to locate a new shopping center, plant a crop, preserve a marsh, develop a windmill, or place solar panels on building rooftops.

For example, the relevant criteria for siting a new park might include 1) a vacant parcel of land at least one acre in size; 2) proximity to the river; 3) a location not too close to an existing park; 4) an area with mature trees; and 5) a location near the homes and work of many people. ArcGIS can readily model suitability for parks and other sites using raster data and imagery. Here are some more examples.

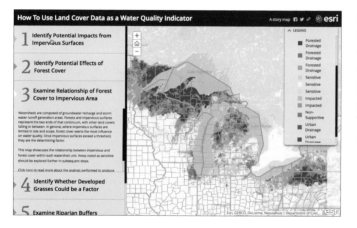

Land cover as a water quality indicator
NOAA has used land cover as a way to predict water quality using GIS analysis. For example, water quality is typically higher in the vicinity of forests and wetlands, and typically lower in regions with industrial facilities and large parking lots. This story map offers an excellent overview of the approach.

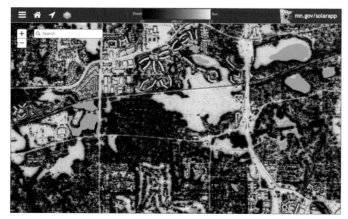

Calculating rooftop solar potential
The state of Minnesota modeled solar potential for the whole state by deriving solar radiation and aspect from elevation, vegetation, and other critical raster and imagery layers. This enables citizens to perform a quick, high-level assessment of where solar power might be a practical alternative for their locations.

What can I see?
Visibility analysis

Viewshed analysis involves analyzing what is or isn't visible from a given location based on distance, terrain, and even land cover. It is an operation that enables you to identify the locations from which a particular landmark is visible; for example, from which areas in a park can I see a river, or how many windmills are visible from the town square?

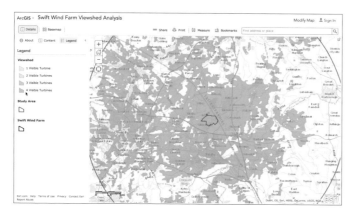

Placing wind farms appropriately

This viewshed analysis determined the visual impact of a wind farm with four large turbines in a study area in England. The visual impact of building a wind farm in urban or semiurbanized areas has the potential to create controversy in the community. Being able to show from where a turbine farm would be visible before one is built helps utilities mitigate reaction.

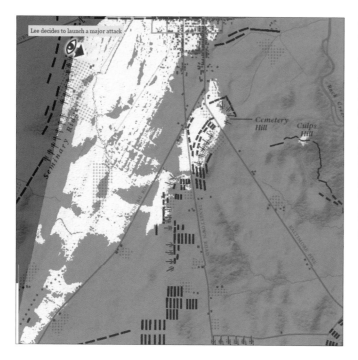

Calculating viewshed

This interesting story map uses GIS visibility analysis to tell the fateful tale of the Battle of Gettysburg in the American Civil War. At the moment General Robert E. Lee (at the red eye) committed to engage with Union troops, he could only see the troops in the light areas; everything shaded in gray (the much greater part of the Union's strength) was invisible to him at that moment. Historians using personal accounts, maps of the battle, and a basic elevation layer were able to unlock the mystery of why Lee may have committed to battle facing such poor odds.

Following the flow of water
Hydrological analysis

Hydrology is the science concerned with the earth's water, especially its movement in relation to land. Because water moves in response to gravity, the elevation of the earth's surface can be used to model how water moves.

Modeling flash flood events

Flood-prone canyons pose a significant threat to recreational users in the semiarid western United States. The NOAA National Weather Service Forecast Office in San Diego has recognized the flash flood risks that exist and has implemented enhanced flash flood services for two flood-prone canyons. This story map details the methods used to create public awareness of the highest-risk areas in the Anza-Borrego Desert State Park.

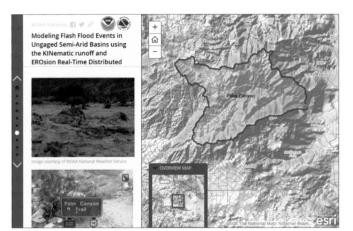

Flooding frequency

Watershed analysis layers provide an estimate of flood frequency as one of six classes from none to very frequent. Click any spot on the map to get a readout of the flooding frequency. This 30-meter resolution layer covers most of the continental United States, including Alaska, Hawaii, Puerto Rico, the US Virgin Islands, and several Pacific Islands including Guam and Saipan.

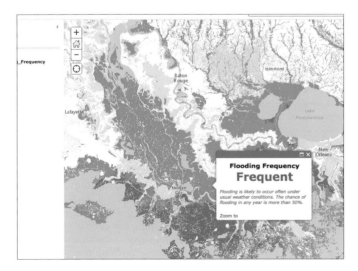

Visualization of rasters
Renderers, 2D and 3D

Raster data can be single band or multiple band, with only a few unique pixel values or with a full range of values in the given pixel depth. And there are a number of ways to visualize raster data as multiband imagery, in 3D and as dynamic time-series maps. For example, when viewing color aerial photography, you are often viewing a three-band raster dataset with an RGB (red, green, blue) renderer applied by default.

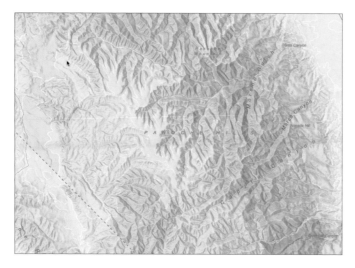

Calculating hillshade
Analytical hillshading computes surface illumination as a raster surface with values from 0 to 255 based on a given compass direction to the sun (azimuth) and a certain altitude above the horizon. Terrain modeling and visualization helps to bring other information layers to life as shown in this map of soils in the Panoche Hills of California, west of Fresno.

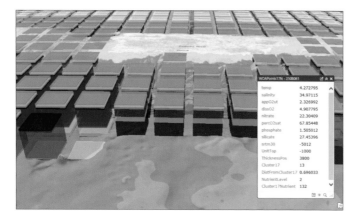

4D visualization of the marine environment
The Coastal and Marine Ecological Classification Standard provides a comprehensive framework for organizing information about coasts and oceans and their living systems. This four-dimensional time series map includes the physical, biological, and chemical properties that are collectively used to define coastal and marine ecosystems. When presented in 3D, the data forms a stack through which an analyst can drill.

Thought Leader: Gerry Kinn

Image analytics is not just about making a pretty picture

Image analysis has evolved dramatically since the first Landsat was launched. Initially, the emphasis was on image processing to make the imagery interpretable; later, to extract features which were used to populate GIS databases. Now, much of the required technology is commonplace. The new emphasis is on processing the imagery in ways that enrich our understanding of the world so that we can better forecast and manage what is about to happen and get ahead of the curve. This is what we are trying to achieve in agriculture, forestry, environmental resource management, urban planning, traffic management, and even in fields like law enforcement.

Image analysis doesn't evolve in a vacuum. It is influenced by related things that are evolving. Today and into the future we see increased computing power with the parallel capabilities of cloud computing; we see more imagery from more modalities with better resolution and more collection options; we have access to massive existing GIS data collections before we even collect the image; and we have new and innovative ways to perform analysis. So where does this perfect storm of progress take us?

Let's take agriculture as an example. In the United States, the vast majority of growers register their fields and crops with the USDA. There are good soils maps and elevation models are available for the entire country. With NEXRAD there is rainfall data from ground-based radar that is collected for the continent every five minutes for cell sizes of less than a kilometer. Combined with other temperature data and daily solar illumination data, crop models can predict what the state of growth

Gerry Kinn is helping to integrate rich imagery technology into the ArcGIS platform so that the near real-time information can help users answer important questions.

should be for every field. This allows for the use of multispectral imagery to validate and adjust these crop models whenever new imagery is collected, whether it be from satellites, aircraft, or drones. In fact, drones make the data very personal for the individual grower, offering better resolution and collection frequency. The analytical results show anomalies where the grower should take action to mitigate moisture issues, nutrient deficiencies, or weed and pest pressures. The net result is better understanding of production whether it is at the national level or for precision agriculture at the field level.

Image analytics are no longer just about making a pretty picture. Rather, they combine the science of remote sensing with all the other available sensor and GIS data to model the important processes that occur every day in our landscapes and affect our lives.

Imagery is the global monitor

What can my map show me?

The goal of cartography or any style of information design is to highlight what is significant about the data. In many cases, when we let the data come to the surface, it is a sophisticated spatial analysis that becomes the map, or the information display.

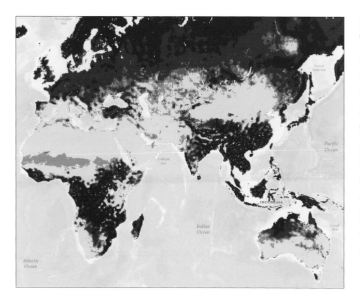

Precipitation runoff index

When precipitation falls on the surface of the earth, much of it is captured in storage (such as in lakes, aquifers, soil moisture, snowpack, and vegetation, among others). Precipitation that exceeds the storage capacity of the landscape becomes runoff, which flows into river systems. In urban areas, pavement and other impervious surfaces drastically increase the amount of surface runoff, which sweeps trash and urban debris into waterways and increases pollution and severity of floods. In agricultural areas, surface and subsurface runoff can carry excess salts and nutrients, especially nitrogen and phosphorus.

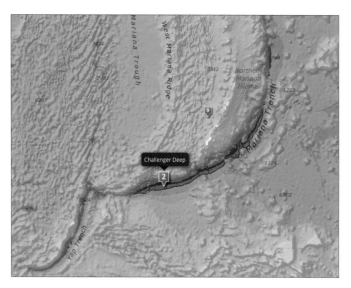

Bathymetry

Bathymetry is the study of underwater depths of lake beds or ocean floors. In other words, bathymetry is the underwater equivalent to topography. This map explores the world's oceans and their bathymetric features.

Analysis case study: Hurricane Irene
Using lidar imagery to model hurricane damage and erosion repair

When Hurricane Irene struck the Outer Banks of North Carolina in 2011, the storm surge and winds carved two new channels through Pea Island. The main transit route back to the mainland was destroyed. Lidar and imagery were flown by state and regional transportation agencies to collect multispectral data and surface information.

The damaged road was the only way in and out for local residents. Not only did the roadway itself need repair, but the surrounding beach also had to be rebuilt as a buffer zone to protect the new road. As soon as the imagery was flown and analyzed (mere days after the event), it was made available to responding agencies and proved invaluable in getting the infrastructure rebuilt.

The state of North Carolina deployed a simple app that allowed officials to begin making calculations about how many truckloads of sand would be required to replace all that had been washed away by the storm. By drawing different-sized shapes on the ground, they were able to provide some realistic estimates for how much sand was needed to get the roadway and beach repaired as rapidly as possible.

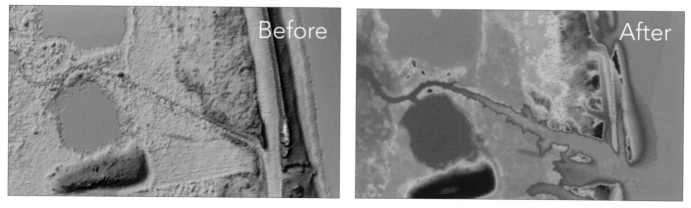

In these dramatic images, the extent to which the sea encroached onto the community (and the amount of sand it took with it on the way out) is made obvious by lidar imagery. So accurate were the measurements, officials were able to calculate how many dump trucks of sand would be required to breach the gap and begin rebuilding the roadway.

In this map of impervious surfaces, we can see how buildings and pavement on each parcel inhibit water from soaking into the ground. Property owners are assessed runoff fees calibrated to the amount of impervious surface on their lands.

Calculating impervious surfaces
Using raster functions

Many local governments use the amount of impervious surfaces to calculate the storm water bill for individual properties. Using dynamic image processing, impervious surface features are extracted from multispectral imagery and then used to compute the total square footage of impervious surface per parcel, as shown in this example from Charlotte, North Carolina. This analytical calculation provides an excellent illustration of the synergy that is enabled by the integration of GIS and image processing.

This subdivision has plenty of asphalt and other impervious surfaces, but it is difficult to quantify.

"Training" is the technical name for the task of identifying which sample segments represent which type of coverage, leading to more accurate automated classification.

Infrared is useful for detecting and extracting vegetation. Red is important for discriminating bare soil. Blue is important for discriminating urban features, especially concrete and rooftops.

Quickstart

Discover new insights, knowledge, and understanding with spatial analysis

▸ Spatial Analyst

ArcGIS Spatial Analyst is an extension of ArcMap that augments the capabilities of ArcGIS for Desktop by adding a range of raster spatial modeling and analysis tools. It is used to solve complex problems such as optimally locating new retail stores or determining the most promising areas for wildlife conservation efforts. While beyond the scope of this book, it's an important tool in the serious analyst's kit.

▸ Going places with Spatial Analysis MOOC

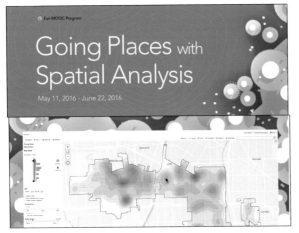

This course is for people who know something about data analysis and want to learn how the special capabilities of spatial data analysis provide deeper understanding. You'll get free access to the full analytical capabilities of ArcGIS Online, Esri's cloud-based GIS platform. Previous experience with GIS software is helpful but not necessary.

▸ Analysis tools in ArcGIS Online

While the Spatial Analyst extension has the largest assortment of raster-specific tools, the ArcGIS Online web GIS environment also has a growing number of such tools. An overview of some of the raster modeling tools in ArcGIS Online is found below.

Calculate Density

The Calculate Density tool creates a density map from point or line features by spreading known quantities of some phenomenon (represented as attributes of the points or lines) across the map. The result is a layer of areas classified from least dense to most dense.

Find Hot Spots

The Find Hot Spots tool will determine if there is any statistically significant clustering in the spatial pattern of your data.

Interpolate Points

The Interpolate Points tool allows you to predict values at new locations based on measurements from a collection of points. The tool takes point data with values at each point and returns areas classified by predicted values.

Create Buffers

A buffer is an area that covers a given distance from a point, line, or area feature.

Learn ArcGIS Lesson

Calculate Impervious Surfaces from Spectral Imagery

▸ Overview

Ground surfaces that are impenetrable to water, known as impervious surfaces, present serious environmental problems. Runoff from storm water can cause flooding and carry contaminated materials into lakes and rivers. Due to these hazards, many local governments enact fees on land parcels with high amounts of impervious surfaces. Among these is the local government of Louisville, Kentucky. However, to place a storm water bill on properties, they need to know the area of impervious surfaces contained in each land parcel.

You'll help them by calculating the impervious surfaces of a single Louisville neighborhood. With the aid of an ArcGIS Pro task, you'll extract bands from a multispectral image of the neighborhood to emphasize urban features like roads and gray roofs. Then, you'll segment and classify the image into land use types, which you can reclassify into either pervious or impervious surfaces. After assessing the accuracy of your classification, you'll calculate the area of impervious surfaces per land parcel to supply Louisville with the necessary information to determine storm water fees.

▸ Build skills in these areas:

- Following a workflow with an ArcGIS Pro task
- Performing a supervised classification
- Assessing classification accuracy
- Calculating land-use area per feature

▸ What you need:

- ArcGIS Pro 1.2.0 or later
- ArcMap 10.3 or later (optional)
- Estimated time: 1 hour 30 minutes

Start Lesson

Esri.com/imagerybook/Chapter5_Lesson

Creating Mirror Worlds
Enabling a new dimension with 3D imagery

When imagery and 3D come together, you're approaching a representation of reality that has been described as a "mirror world." On such a virtual Earth, remotely acquired imagery provides the foundation for accurate modeling of the true shape and texture of the world. The mirror world has evolved into an accurate 3D map, not just a fancy model, and that's a big deal because it opens the door for things like true 3D feature extraction; volumetric analysis; and interactive, reality-based earth visualizations from a global scale all the way down into building interiors.

The evolution of 3D imagery

Prior to the 1990s, mainstream 3D imagery was more about transforming real-world measurements of the terrain into 2D digital representations of the earth's topography. This data was viewed in 2D given the general lack of 3D viewers and enormous computing requirements needed to work in 3D.

From about 1990 to 2010, 3D software gradually became more accessible and computing power followed Moore's law with increasing performance and falling prices. Three-dimensional objects like buildings began to model the real world in a GIS. Adoption of 3D GIS was still not widespread, and most 3D work was still performed by draping two-dimensional maps on 3D representations of the terrain. Three-dimensional objects were created in labor-intensive processes by skilled data operators using stereographic image pairs. The ability to create 3D objects was still outside the grasp of most users.

Fast-forward to today's technology. We now have an end-to-end workflow for processing 3D imagery; creating, editing, and maintaining 3D elevation from sensors and imagery; conducting analysis and visualizations; and sharing immersive 3D scenes to desktops, browsers, and mobile devices. Never before have these tasks been as readily accessible. Using ArcGIS on desktops, your enterprise server, the web, and mobile devices, you can use imagery to fully 3D-enable your GIS. Your work may involve topics as varied as global climate change; regional forest management planning; creating sustainable, resilient, livable cities; or all the way down to site-specific work on a park or inside a building.

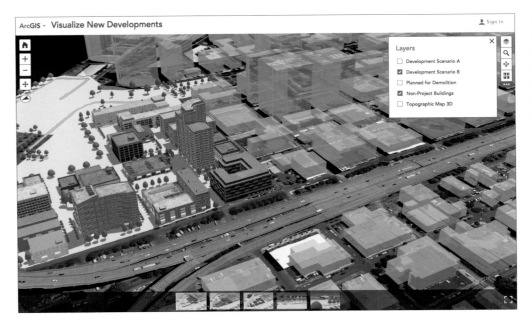

Comparing proposed urban developments in Portland, Oregon, within the context of the existing built environment. The scene includes 3D terrain, extruded nearby buildings, and procedurally drawn 3D buildings. Visualizations like this used to take skilled artists many hours to create; now it can be done by one person in minutes.

3D is embedded in ArcGIS

ArcGIS enables you to immerse yourself in your 3D world, starting with imagery and your GIS data. This can be a photorealistic experience or a way to interact with your analytic results. Whether you're exploring a project site from multiple view angles, publishing a 3D web scene for a website, or creating new 3D layers from scratch, ArcGIS makes it easy to accomplish these tasks.

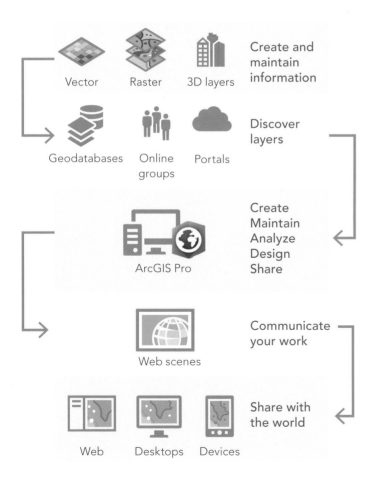

Vector Raster 3D layers Create and maintain information

Geodatabases Online groups Portals Discover layers

ArcGIS Pro Create Maintain Analyze Design Share

Web scenes Communicate your work

Web Desktops Devices Share with the world

Tell a global-scale story about hurricanes traveling across the Atlantic Ocean, route viewers through the interiors of buildings, or go below the surface to look at underground utility features.

ArcGIS allows you to interoperate across devices and platforms. Share your 3D scenes across desktops and on the web, providing the appropriate level of detail and complexity for specific audiences from the general public all the way to professional planners and engineers. New open standard, server-based, 3D streaming technology allows you to share a single 3D building—or millions—via high-performance and easy-to-use web scenes.

ArcGIS Pro is the 3D workhorse of the platform. It runs on your local desktop and provides an intuitive environment in which to create and maintain your 2D and 3D layers. With its powerful analysis tools, you can conduct uniquely three-dimensional studies: evaluation of rooftop solar potential; shadow impact studies on the winter and summer solstice; and accurate valuation of real estate by analyzing the views from different places within a building.

For the highest impact, you can create real-world 3D representations of your ideas and designs by using rule-based, procedural modeling. For example, you can envisage future coastal cities and regions where designing with nature involves developing artificial reefs.

All of this 3D content is stored and easily shared and put to use with ArcGIS Online.

The scales of 3D

With 2D maps, both the perspective and scale are fixed in a top-down view and can be described by the numerical relationship between map units and real-world units (for example, 1:1,000). In a 3D world, elevation and imagery combine in more unique and complex ways—and a single view can contain several different scales (distances, really) all at once. Features in the foreground are closer and thus larger scale than features farther back in your viewing window.

In 3D, we often use broad categories to describe scales. These categories, from the planetwide global view down to the very local site scale, are driven more by the common data patterns and types of work that are done, rather than relying on an exact numerical relationship between map units and real-world units.

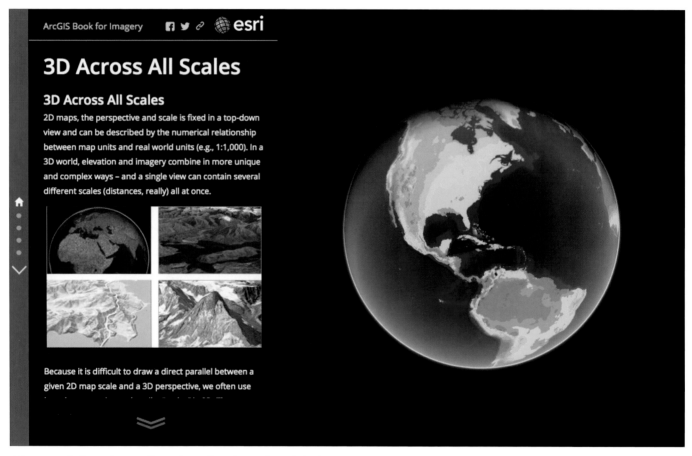

3D in ArcGIS works at all earth scales, from the global perspective into continental and regional views, to cities and neighborhoods, and inside of individual buildings and site locations.

Visualize from a global view to your rooftop

Global

At the global scale, 3D scenes are used to understand or provide context for intercontinental phenomena that wrap around the spherical surface of the earth. Some common examples of datasets with large extents are climate, weather, oceans, global transportation, and presentations about countries and continents. At this scale, the elevation surface is not discernible and imagery resolution can be relatively coarse.

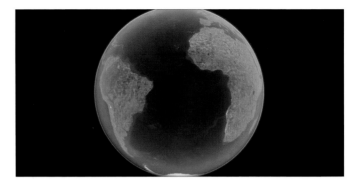

Regional

At the regional scale, which includes counties and provinces up to country-sized areas, we start to detect some of the more dramatic changes in the terrain surface, such as canyons and mountain ranges. Datasets can still cover large extents but are observable in greater detail and resolution.

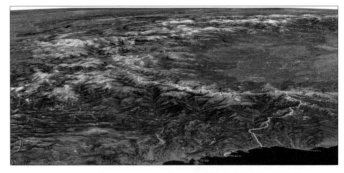

Local

At the local scale, 3D features such as buildings become visible and there is a clear definition between the textures and extents of the built and natural environments. This scale is often used for urban planning and visualization as well as for agriculture and farm management. These scales are enhanced by high-resolution imagery. Detailed, local elevation surfaces are often added, providing more accurate alignment of 3D data with the terrain.

Site

The site scale is used to visualize datasets that cover the extent of a collection of buildings, like a city block, a campus, a community park, or a stand of trees. It can also include indoor maps in 3D. At this scale, having accurate, high-resolution elevation data is critical, and high-resolution imagery is often used. Custom basemaps may also be added that enable zooming into individual rooms and floors inside of buildings.

Capturing elevation with lidar and radar

The earth's topography forms the natural foundation for working in 3D, where objects are placed above, on, or below the terrain surface. In a GIS, the topography is represented using elevation models in raster format or sometimes a triangular irregular network (TIN) structure. Raster elevation models (far more commonly used) divide land surface into a regularly spaced grid, with each grid cell containing a single elevation value. By contrast, TINs rely upon variably sized triangles (or facets) connecting a series of 3D point locations in space, enabling unique elevation values to be interpolated at any location within the facet.

Today elevation models are more commonly created using remote sensing techniques involving lasers and radar. Interferometric synthetic aperture radar (IFSAR) uses pairs of opposing radars to collect elevation readings. An example is the 2001 NASA Shuttle Radar Terrain Mapping (SRTM) mission, which mapped elevation for the entire globe at 30-meter resolution using radar.

Lidar has rapidly been adopted as one of the standards for how high-resolution elevation models are created. This process uses timed laser pulses to produce a highly accurate and dense "point cloud" of elevation points and can produce a number of information products including digital surface models (DSMs), digital terrain models (DTMs), and triangular meshes (known as TINs).

Lidar is active remote sensing that uses lasers to strike features (like Grand Coulee Dam) and record the reflected pulses to generate the 3D model of objects. In this case the data is symbolized by LAS class (water, cement, transmission lines, and more).

An example of detailed lidar collected for parts of Petaluma, California, by the GIS team in Sonoma County.

The foundation for 3D GIS
Base elevation

One of the key 3D attributes of a feature is its vertical position, or base elevation. This is where the base of a building, tree, or other object aligns with the elevation of the so-called "bare earth" surface. Good alignment with the terrain is important from a visual perspective to limit "floating" or "buried" features that should be positioned flush to the earth's surface, but also to ensure the accuracy of relative vertical positioning for 3D analysis.

A bare-earth digital terrain model (DTM) is usually sampled to determine the correct z-values to apply to a feature, be it a point, line, polygon, or multipatch (volume). A single z-value can be applied to vertically shift a feature's location up or down, or the feature can be "draped" onto the elevation surface.

DEM

For practical purposes this bare earth digital elevation model (DEM) is generally synonymous with a digital terrain model (DTM). Quality DEM products are measured by how detailed the elevation is (in other words, the ground size of each pixel) and how accurately the morphology is presented—that is, its z-axis accuracy.

DSM

A digital surface model (DSM) represents the height elevations of the surface trees, buildings, and other features projecting above the bare earth.

The foundation for 3D GIS
The key elevation models

DSM: Digital surface model

Surface models include topography and all objects on the earth's surface, like trees and buildings. Drape imagery over a DSM to create simple virtual worlds, or use specialized tools to create new tree or building features in your GIS. Lidar is typically used to create DSMs working with what is commonly referred to as first return data, as the elevation of the first returned laser pulse is used.

DTM: Digital terrain model

Terrain models, commonly referred to as bare earth, are void of things like buildings and trees. Use a DTM to create hillshades, determine slope of the topography or the aspect to the sun, calculate surface water flow, or set the base height of buildings and other features.

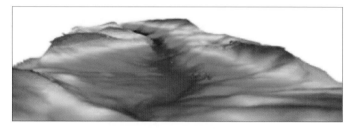

DHM: Digital height model

Less common but critical for 3D-enabling your GIS, height models are used to calculate height above ground for buildings, trees, and other features. Height models are created by calculating the difference between the terrain and surface models. Lidar is increasingly the way that DHMs are created, as in this profile of canopy heights in an old-growth Douglas fir forest.

Bathymetry

The topography of the ocean floor or a lake bed is called bathymetry and can be used in many of the same ways as land-based topographic data. Bathymetric data is generally collected using sonar, which, being an active sensor, has similarities to lidar but is conducted from a ship floating on the water's surface. Some shallow coastal bathymetry can be collected from aircraft.

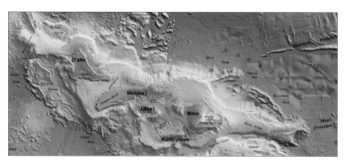

The web scene
3D-enabling GIS data

In ArcGIS, 3D maps are called scenes, and those that work on the web and in browsers are called web scenes. Start with your terrain as the foundation. All 3D scenes start with a surface elevation coverage onto which you can drape any number of 2D maps. The simplest and fastest way to "go 3D" is to drape your imagery or basemaps onto your elevation surface for your base. It is a real revelation to many GIS users when they recognize how many of their GIS layers are already 3D ready and that other layers require just a few small steps to enable them for 3D use.

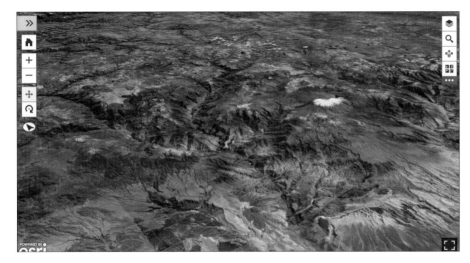

3D displays of imagery provide ready-to-use, photorealistic web scenes for your foundation. It's easy to add additional 2D and 3D layers onto your web scene to bring your immersive 3D GIS experiences to life.

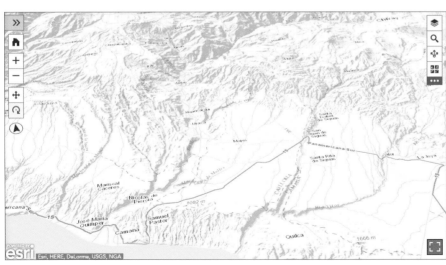

With elevation as your foundation, virtually any 2D map layers can be used to create engaging 3D scenes and experiences for your users. This foundation is just the beginning of what your 3D mapping can represent.

Create an accurate foundation
Collect high-resolution elevation data

When your work requires higher-resolution elevation data for use at larger, zoomed-in scales, two relatively new sources can provide the added extra detail. These include true lidar collections and drone missions, both of which can be used to generate high-resolution surface elevation data. Collection of high-resolution lidar is growing rapidly, enabling 3D GIS for whole new applications.

Drone flights can also be used to generate elevation values for every photo pixel from the photos being collected. This generates a digital surface model of the height of all observed features, not just the bare-earth elevation readings.

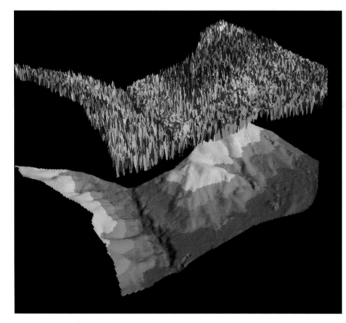

In this image we see the separation of elevation of the vegetation biomass from the bare-earth surface. Both biomass and surface elevation are captured as part of the first return of the signal, while the bare earth is calculated by analyzing the lidar returns to extract the DTM surface elevation.

Collected with a consumer-grade drone and processed through Drone2Map, this side-by-side view shows what the digital surface model looks like compared to a straight-down orthomosaic.

Create your 3D basemap

ArcGIS comes with global elevation sources that provide 30-meter-resolution elevation for most of the world. If you have a higher-resolution elevation source like lidar or drone results, you can create your own 3D large-scale mapping layers by adding your detailed elevation for use when zoomed into your areas of interest.

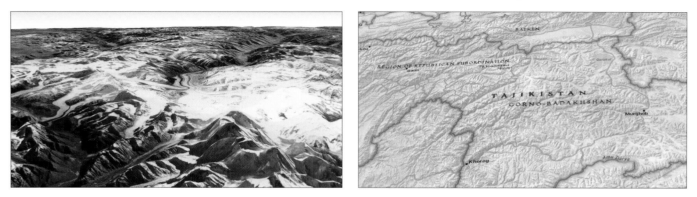

Any data draped onto a three-dimensional surface takes on new perspectives. Whether it's image data from the Himalayas, or the National Geographic style basemap, raster data finds new life in 3D.

Transform 2D features into 3D

Some 2D GIS features can be creatively represented and used in 3D GIS by including just a few extra attributes for each 2D feature.

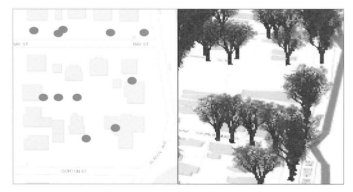

Tree locations can be collected as point features. With a few simple, added attributes—like species or common name, crown height, and width—you can symbolize realistic properties to represent the tree from a catalog of tree symbols that can be sized and placed at each tree's point location.

Buildings can be extruded based on their height or number of floors. And you can model them with different levels of detail depending on your needs. The 2D building footprint (left) can be extruded as a block building (center) or modeled by adding surface detail (right).

Modeling buildings and tree heights

The height of a 3D feature is determined by measuring the distance between its intersection with the ground surface and its highest point. With this basic z-value information, a 2D feature can be extruded, or an existing 3D model can be stretched to the correct height.

Using classified lidar data, the resulting digital terrain model (DTM) can be subtracted from the digital surface model (DSM) to create a "height above ground" raster, or a normalized digital surface model (nDSM). This nDSM raster is then sampled within the 2D or 3D GIS feature to determine the maximum height value, which is applied to the geometry via extrusion or vertical scaling.

Buildings

Building footprints are first aligned with the terrain surface; then extruded by the height value; and finally textured according to any attribute, for example, the number of floors.

Trees

Trees are another feature commonly extracted from imagery or lidar. Using the same procedures for determining the base elevation and maximum height of a building, tree heights can be extracted using either manual or automated methods. Detailed 3D symbols can then be applied according to the tree species (for example, oak trees), and then scaled vertically by height and laterally by crown width of each tree to create realistic symbols.

Realistic 3D tree models are scaled by deriving height and crown width attributes from lidar.

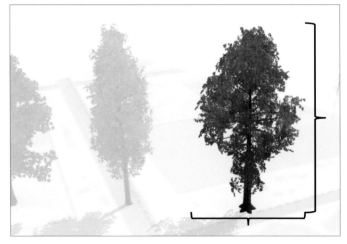

Representing 3D buildings

3D buildings can be represented within a range of detail or accuracy, which is usually determined by the methods or source data used to collect them. Popularized by CityGML, a schema for representing virtual 3D city models, these varying levels of detail (LoDs) show different degrees of abstraction from the real-life 3D structure. Each LoD can exist in both a textured and untextured form, and choosing the right level of detail depends on the availability of data to extract the building geometry and the specific use case for 3D visualization and analysis. For example, a LoD1 building is not appropriate for conducting shadow impact or rooftop solar analysis due to the lack of accurate roof shape. Likewise, a realistically textured LoD4 building contains more detail than is needed for exterior 3D analysis.

LoD0—A level of detail 0 building is simply a 2D footprint or polygon with base elevation information. In other words, it is vertically aligned in some way with the terrain surface.

LoD1—An LoD1 building is a polygon that has been extruded to a given height, resulting in a closed 3D shell that consists of only horizontal and vertical planes. LoD1 buildings are easy to create and show relative heights, but due to their simplified geometry, they are best suited to 3D visualization rather than 3D analysis.

LoD2—An LoD2 building shell includes vertical wall surfaces, in addition to roof form geometry, and may include details such as chimneys or dormers. LoD2 buildings are appropriate for a variety of 3D analyses, like shadow impact, line-of-sight, and rooftop solar potential.

LoD3—An LoD3 building shell includes detailed wall and roof geometry as seen in LoD2 buildings, but with additional fine architectural details like windows, doors, or columns. LoD3 buildings are good for street-level visualization and analysis of focused sites, where up-close realism matters.

LoD4—Building shells include not only detailed exterior features but also interior walls, floors, doors, and furnishings. LoD4 buildings are appropriate for indoor GIS and visualizations.

Extracting 3D roof forms

Creating true-to-life 3D building forms has historically been outside the technical capabilities of most municipalities and urban planners—the GIS users who could benefit most. Creating this data was often prohibitively expensive. Fortunately, as more urban areas are being surveyed with high-resolution lidar and photogrammetry, detailed 3D information has become more accessible. From this data we can extract the information necessary for modeling buildings not only with accurate height information, but with accurate roof forms as well.

The key to modeling a building procedurally is accurately representing the building's roof geometry. The attributes necessary to capture this geometry are the following:

Total height—Height of the highest point of the building (that is, the roof's ridge).

Eave height—Height of the main eaves of the building roof form.

Roof type—Type of roof structure (flat, gable, hip, and so on).

Roof direction—Direction of the roof's ridge.

While these attributes can be entered manually by a technician examining each building one at a time, automated procedures can be used to extract roof form information based on standard tools in ArcGIS. This process is described in the Learn ArcGIS lesson at the end of this chapter (page 112).

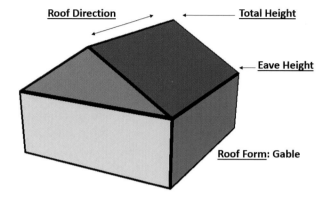

The basic attributes for procedurally creating a 3D building.

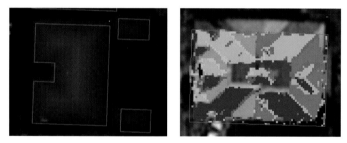

Building footprint polygons on a digitial surface model raster (left) and a rooftop classified by slope and aspect.

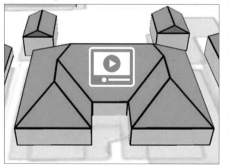

Watch a video: See the roof form extraction tool at work.

Thought Leader: Pascal Mueller
Mapping the future of our cities

More than half of the world's population now lives in cities, and the percentage is growing. While the size of the rural population will be roughly the same as today, the total numbers of urban inhabitants will have doubled to seven billion by 2050. As a consequence, cities will grow significantly larger in area and taller in height.

Three-dimensional GIS is the key tool for shaping this urban future. An urban space is a complex collection of three-dimensional architectural structures arranged into living units, buildings, parcels, blocks, and neighborhoods. These structures are interconnected by different types of transportation networks such as roads, pedestrian routes, and metro lines, and are interfacing with multiple layers of energy and utilities networks. A 3D GIS such as ArcGIS provides numerous tools for modeling and managing these existing structures. This results in so-called smart cities where all departments are connected with each other and the performance of the infrastructure as well as the quality of the urban services are monitored in real time.

Traditionally, modeling and planning urban spaces in 3D has been mostly a manual task that consumes significant amounts of resources. But with the growing requirements of quantity and quality in urban 3D content, there is an imperative need for alternative solutions that allow for fast, semiautomatic urban modeling, design, and simulation. Examples include the visualization of 3D building shells out of 2D footprints; maintaining 3D cadastral records of apartments in skyscrapers; managing zoning regulations in 3D and studying the economic impact of zoning amendments; and master planning where the performance of urban redevelopment—typically densification scenarios—is simulated, analyzed, and optimized in regard to space demands, traffic impacts, sustainable energy usage, and quality of living.

Three-dimensional capabilities are available across ArcGIS, allowing smart cities also to engage more effectively and actively with their citizens. Therefore, 3D information can now be shared easily with the public to better communicate design decisions and visualize big data in its urban 3D contexts.

Pascal Mueller is the Director of the Esri R&D Center Zurich, where 3D software for ArcGIS as well as the Academy Award–nominated procedural city modeling tool CityEngine are developed.

 Watch Pascal and the rest of the CityEngine development team

Texturing

Applying imagery to buildings or other features can transform blank 3D geometries into realistic-looking, detailed 3D models. Pasting aerial and low-angle oblique (or street-level) imagery to individual building faces can be done in one of two ways: realistically or procedurally.

Realistic textures
Realistic textures are generated from aerial or oblique images captured for the actual building or feature that is being modeled in 3D. Every surface of a building has a unique texture that represents what each individual facade or roof surface looks like in real life. This approach typically involves higher data acquisition costs and labor, but creates a very realistic 3D scene; for example, familiar or iconic buildings look as they should, with all the correct architectural details.

Photo textured, realistic Mont Blanc in the French Alps.

Accurately textured, realistic 3D building models for downtown Indianapolis, Indiana.

Procedural textures

Procedural texturing involves the application of a geotypical or architecturally typical facade and roof textures onto 3D building forms. These textures can be applied to existing, untextured 3D building models, or they can be applied as part of the procedural modeling process of converting 2D footprints into textured 3D buildings.

Procedural textures are applied according to rules that determine the type of texture for each building, and how many times it is repeated. These textures are typically collected from oblique or street-level images, which are then rectified and edited so they can be repeated seamlessly around the perimeter of each building or floor.

To create the most realistic-looking 3D buildings, texture libraries are grouped and can be applied to match the land-use type of the building, its height, the number of floors, and the regional architectural style. Often, texture libraries are split into two groups: the ground floor, which is usually taller and may contain entrance doors and store frontages, and the upper floors, which repeat uniformly.

While the textures on a procedurally textured building may not exactly match their real-life counterparts, they are inexpensive to generate and closely approximate the look and feel of different general building styles. Procedural buildings are also very useful in urban planning, where the heights and styles of proposed buildings can be changed easily to evaluate different scenarios.

Procedurally textured 3D buildings for downtown Greenville, South Carolina.

Collecting 3D data
How drones capture data in three dimensions

One of the great miracles of the current technological era is the ability to generate a 3D map of a small study area using drone imagery and Drone2Map, a new application from Esri. In the basic building inspection workflow depicted here, the drone flies a preprogrammed path and collects a series of oblique-angle images. The software then does the work of creating a 3D scene.

1. Fly the mission

To inspect this building, the drone flew a near perfect circle (yellow). The capture points are the blue dots.

2. Capture oblique photos

Shown here are 4 of the 40 total oblique-angle images captured by the drone's high-resolution camera.

3. Create the 3D scene

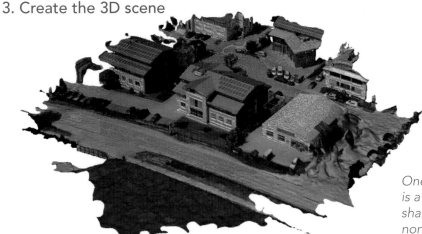

One of the available outputs from Drone2Map is a 3D PDF, which is an excellent means of sharing 3D scenes in a reliable way with non-GIS user audiences.

Case study: The Oatlands House mission

Two centuries ago, George Carter designed and built Oatlands Historic House and Gardens in Leesburg, Virginia. Today, the property is maintained as a historic preservation site. Flying drone imagery, the property managers were able to quickly gather a high volume of accurate data, including a detailed 3D model that could be used to visualize proposed landscape changes (such as large tree removal).

3D scene

Photo

This 3D web scene was generated from a point cloud collected by a consumer-level, personal drone. This scene was available online in under 30 minutes, once the drone mission was completed.

Watch a video presentation about the mission

Mesh

A mesh is a data structure that partitions geographic space into contiguous, nonoverlapping triangles that can be painted with RGB values.

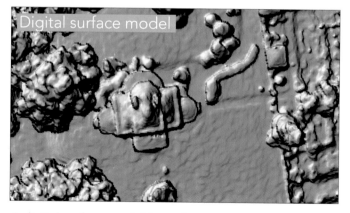

Digital surface model

A digital surface model from the same site reveals an impressive payload of rich and detailed elevation data.

Quickstart

Some places where imagery is exposed in 3D with ArcGIS

▸ Scene Viewer

The scene viewer is an app built into the ArcGIS Online website for creating and interacting with 3D scenes. The scene viewer works with desktop web browsers that support WebGL, a web technology standard built into most modern browsers for rendering 3D graphics. You can also sign in to make your own scenes.

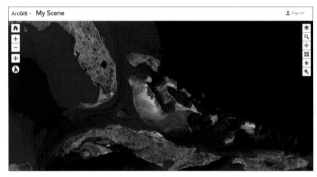

▸ Drone2Map

Drone2Map for ArcGIS is a desktop app that turns raw still imagery from drones into orthomosaics, 3D meshes, tile images, and more, in ArcGIS. Create 2D and 3D maps of hard-to-access features and areas.

▸ ArcGIS Earth

ArcGIS Earth is a freely available, lightweight desktop app that makes viewing 3D maps instant and easy for anyone. ArcGIS Earth is available for Windows desktops and tablets.

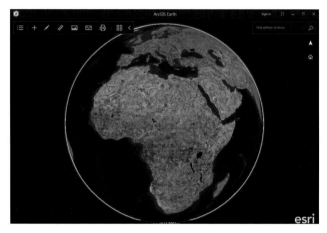

▸ CityEngine

CityEngine is an advanced tool for scenario-driven city design and developing rules for creating procedurally built data.

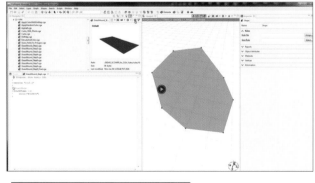

City Engine Quickstart

Learn ArcGIS Lesson

Extract Roof Forms for Municipal Development

▶ Overview

The municipal government of Portland, Oregon, wants to evaluate buildings in a downtown neighborhood to assess whether they follow the city's new Green initiative. The evaluation will include many metrics, such as solar and shade analysis, that require a 3D scene of the area with realistic building roof forms. A basic 3D scene uses level of detail (LoD) 1 buildings: building footprints extruded to a uniform height. The municipal government wants LoD2 buildings, which depict roof form attributes like eaves, gables, and slopes.

Your goal is to create a 3D scene of an area in Portland with LoD2 buildings. With the aid of an ArcGIS Pro task, you'll create a point cloud dataset derived from lidar data and use it to make digital elevation models of the area. Based on patterns in the elevation models, you'll add attribute data about roof forms to building footprints and then symbolize the footprints in 3D with a rule package. Lastly, you'll check the buildings for errors and edit incorrect features before converting the data to a multipatch feature class that you can easily share with the municipal government.

▶ Build skills in these areas:
- Following a workflow with an ArcGIS Pro task
- Creating LAS point cloud datasets from lidar data
- Creating LoD2 buildings for a 3D scene
- Editing 3D features

▶ What you need:
- ArcGIS Pro 1.2.0 or higher
- Estimated time: 1 hour 20 minutes

Start Lesson

Esri.com/imagerybook/Chapter6_Lesson

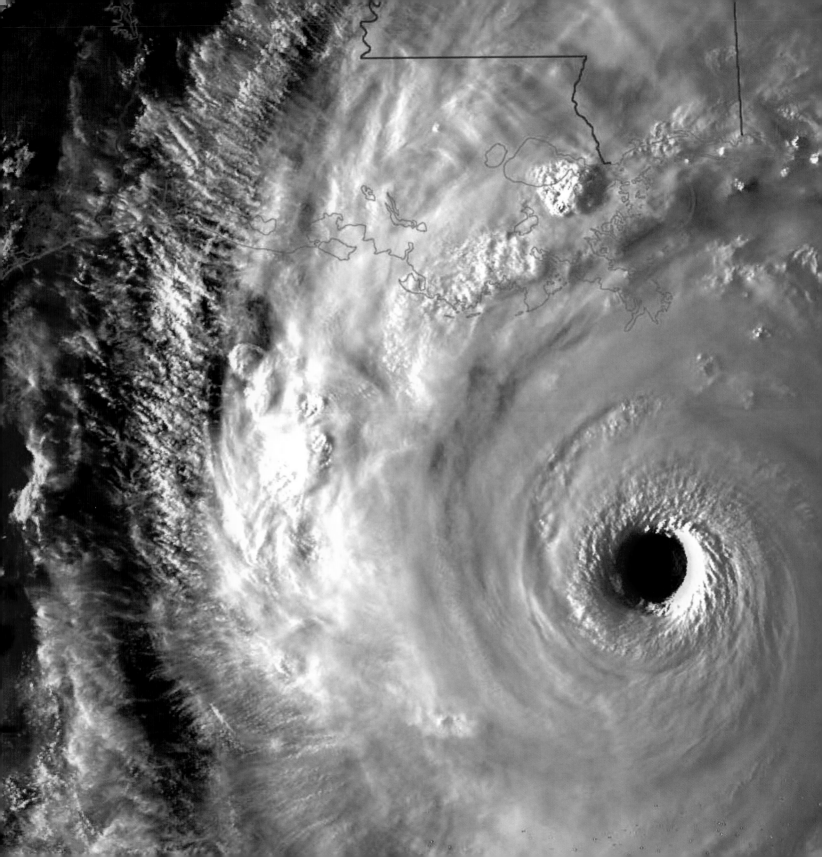

———————— ⬡ 07 ————————

Imagery in the 4th Dimension
The world's greatest time machine

Time series or temporal imagery creates an opportunity to systematically make comparisons over time. The result is a broad array of potential time-traveling applications such as before-and-after views of cataclysmic events and reconstruction of historical landscapes, as well as monitoring and forecasting change over weeks, months, years, and decades. The study of dynamic processes over time is the big idea of this chapter.

It's about time

Using imagery to monitor our dynamic planet

To truly understand our dynamic planet, we strive to explore information through time, visualize the past, understand the present, and recognize future trends. For example, earth scientists use a time series of satellite observations to track monthly precipitation patterns such as snowpack coverage and extent as it descends from the higher latitudes and the poles during the winter months and then recedes in summer. Scientists use satellite time series observations to monitor droughts. And climatologists apply models to forecast climate trends for future points in time. As computing and GIS continue to grow, new time-based capabilities are being developed and applied. Along with this trend, there is a growing appreciation for the critical importance of imagery's temporal aspects. Our world is dynamic so it makes sense that GIS reflects that. And imagery plays a pivotal role.

There are countless applications that require temporal considerations. Delineating a wildfire boundary requires a series of missions to collect thermal imagery for a focused area around the fire perimeter. These are often flown several times a day since wildfires change rapidly. In contrast, exploring deforestation requires imagery covering a large area but for longer time spans, since deforestation is a process that occurs over periods of many years or decades.

Continuous global observation

Meanwhile, there is no lack of raw information coming from sensors to address many of these challenges. There is an ongoing explosion of continuously observing satellite platforms—both government and commercial—all contributing to our collections of earth observations. A number of satellites are designed for continuous observation, revisiting the same areas over repeating time periods—Landsat, MODIS, GLDAS, Sentinel, SPOT, and RapidEye, to name just a few. Landsat 8 revisits every Earth location about every 16 days (its "revisit period"). MODIS collects observations resulting in global coverage every one to two days. And so on.

It's about time.

You perceive temporal transformations as they happen, such as the passing of fall to winter and the shift from day to night. Imagery mimics and extends human perception to much grander temporal and spatial scales. Time-aware imagery empowers us to ask and answer questions that transcend our personal time and space.

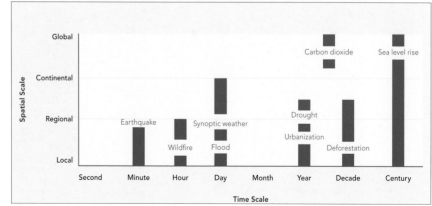

Critical challenges to our planet occur at various space-time scales.

Types of time
Discrete, cyclical, and continuous

Time can be viewed as linear or cyclical. Linear time has a distinct beginning and end and can be expressed using discrete, continuous, or cyclical measures of time. Video is an example of imagery captured in continuous time. Cyclical time captures events that occur in a sequence over and over. Weather that is observed daily is an example of cyclical time.

Continuous

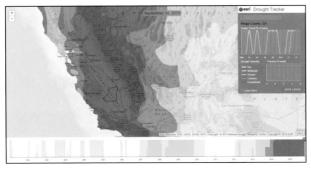

Providing an interactive interface to changing drought conditions in the United States, the Esri Drought Tracker app is an example of detecting changes over linear time. More severe water shortages result in crop damage and require voluntary water-use restrictions. A severe drought can devastate crops and livestock, reducing farmlands to dust. The longer the land goes without restoration from snowmelt and rainfall, the more severe the drought and its possible consequences.

Cyclical

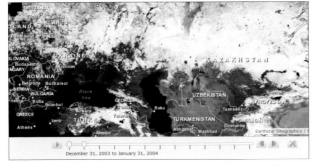

This map features NASA's Blue Marble: Next Generation imagery in a set of 12 monthly composite images of the entire earth, using 500-meter-resolution imagery from the MODIS satellite. These monthly images reveal seasonal changes of the land surface: the green-up and dieback of vegetation in temperate regions such as North America and Europe, dry and wet seasons in the tropics, and advancing and retreating Northern Hemisphere snow cover.

Discrete

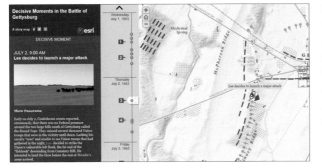

Imagery and GIS-based viewshed analysis explore the viewpoints of Union and Confederate commanders at discrete milestones during the Battle of Gettysburg in the American Civil War. Panoramic landscapes show how what commanders could see had a significant effect on the decisions they made. Click the milestones shown on the vertical timeline to explore each general's actions and the corresponding battlefield conditions.

Historical imagery

Imagery is one of the most effective and moving ways to capture the past. Historical imagery serves as the benchmark for detecting change and allows us to make better decisions in managing the earth's valuable resources. GIS technology brings new life to historical maps and old photographs—on the ground, from the air, and even via outer space.

Historical scanned maps

Scanning paper maps turns them into imagery. Once scanned, they can be georeferenced and included in a GIS just like any other layer. Historical maps can provide context for your analysis and serve as a basis for change detection. The USGS has one of the world's largest collections.

Significant residential development of marshland between 1891 and 1963 in New Orleans is apparent in these scanned, historical USGS topographic maps.

Historical aerial photographs

This type of imagery is taken from the air using balloons; aircraft; and more recently, drones. It provides a planimetric (bird's eye) view of the landscape. The first aerial images were taken in the late nineteenth century.

These images taken in 1970 and 2010 over Dubai show the unprecedented growth that has taken hold of this oil region.

Historical terrestrial photographs

Photography was the first method available for capturing imagery. Historical photographs can provide context and perspective for a GIS analysis. You can locate the same place photographed a hundred years ago and photograph it again for imagery comparison, to see what's changed, as the USGS Repeat Photography Project did along the Colorado River.

USGS rephotography (May 31, 1889, on the left and March 24, 1997, on the right) of Cataract Canyon on the Colorado River shows the proliferation of non-native cheatgrass.

Modeling Earth's processes

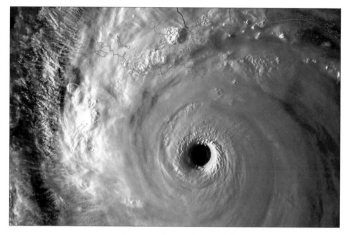

Government agencies such as the USGS and NOAA maintain large, searchable archives of imagery. Historical satellite imagery allows us to monitor the impacts of earthquakes, fires, and severe weather on our planet.

There are a number of Earth observation satellites that pass over selected areas of the planet daily to produce a time series of imagery. For example, weather satellites show the day's weather as it happens in near real-time. Meanwhile, a number of satellites such as GLDAS pass over all global areas daily producing critically important observations about our planet.

Interestingly, multidimensional computer simulation models of Earth's physical processes are used to compute results with a time series of images, often to interpolate what happens in between satellite passes. Other models can also simulate historical conditions to re-create past events as well as forecast future conditions.

This 2005 animation of infrared satellite imagery shows Hurricane Katrina's intensity: white is the most intense, followed by red, green, and blue.

GLDAS soil moisture data is useful for global-scale modeling of water cycle scenarios like drought and flooding, which are currently having huge impacts on not only humans but also widespread biodiversity.

Studying the past

Imagery enables us to digitally reconstruct past landscapes. We can use one historical image to study a location in the past or a collection of historical images to see in more detail how something has changed over space and time.

Spyglass on Chicago before the Great Fire

"What Did Chicago Look Like Before the Great Fire?" by Smithsonian magazine, demonstrates how historical imagery can be used to understand how things change over space and time. Readers of the online edition are treated to this embedded story map that compares the 1868 pocket map of Chicago with the most recent aerial imagery available. Use the spyglass to compare the two maps. Explore the waterfront to see the new land created from the fire debris.

Organizing imagery collections and historical maps through time into mosaics provides an especially useful tool for comparing previous landscapes with each other and with the present.

Water loss in the Aral Sea

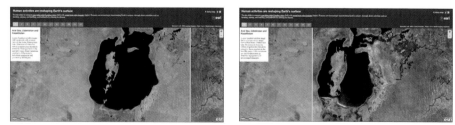

Once one of the largest lakes in the world, the Aral Sea has steadily shrunk since the 1960s due to the diversion of the rivers supplying the lake for cotton irrigation. The 1990 Landsat satellite image (on the left) compared with the 2015 view of this briny lake in Central Asia shows a noticeable loss of water surface area over several decades. Today the lake is only about 10 percent of its original size.

Thought Leader: Greg Allord

Historical maps provide a window to the past and context for the future

For many years at the US Geological Survey (USGS), I was able to support, observe, and greatly respect the work of skilled geologists, hydrologists, engineers, water quality specialists, illustrators, editors, and cartographers. Be it the artistry and precision of the cartographers, the era of print media, the interpretative science, or the geographic extent, the content of the maps and publications that our teams created appealed to a wide variety of people.

Organizations studying the environment, sciences, and culture of the United States have been in existence for many decades, some dating as far back as the first years of this country. The USGS has had a 136-year legacy of geologic, topographic, water, and biologic resources studies. It is also where I had the privilege of working. At the USGS, the transition moved us from manual methods, into a period of digital exploration, to now, with the promise of digital cartography and map sharing having been realized.

Librarians have taught us that it is not sufficient to simply house a hard-copy print collection like this; it must be cataloged and preserved. An effective preservation process is capturing an image of the original. Traditional paper maps and reports, in particular, benefit from this. The result: more than 130,000 USGS science publications can be accessed at http://pubs.er.usgs.gov, and approximately 175,000 topographic maps can be accessed at http://nationalmap.gov/historical/index.html or used in ArcGIS Online. Not only does this make it possible for researchers and

Retired USGS cartographer Gregory Allord led the effort to scan and georeference approximately 175,000 USGS topographic quad maps as GeoPDF files and make them freely available to the USGS and the public online.

scientists to use these maps and documents, but it also allows a broader audience to appreciate and experience their beauty and importance.

The historical maps the USGS and other agencies created provide not just a window into yesterday, but also a context for tomorrow. While it is not possible to predict all users and uses of this information, the accessibility of the archive of historical maps through ArcGIS Online and the Living Atlas provides the world with the ability to better understand the past, manage the present, and plan for and share the future.

View the historic map collection

Historical map collections

Historical maps add an important dimension to GIS. They offer a clarity and shape to what our world was like in the past. They speak of the possibilities of places we can never visit again except through the use of these maps. They offer a framework for comparisons between now and the past as well as into the future. One great thing about historic maps is that they can be integrated with maps and information from today's world. In effect, they can be added as new kinds of layers to your GIS. The way this is done is by scan-digitizing the historic maps and georeferencing them. They essentially become a new kind of raster layer in your GIS and create enormous opportunities for many types of applications.

Africa 1787

This map features a composite of Africa, with all its states, kingdoms, republics, regions, and islands, circa 1787. The famous work shows many interesting place-names and country boundaries that no longer exist.

1942 Theater of War

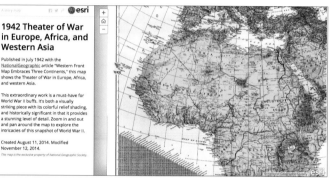

Published in 1942 by National Geographic, this map shows the Theater of War in Europe, Africa, and western Asia. It accompanied the article "Western Front Map Embraces Three Continents."

Pittsburgh Map and Historical Site Viewer

This site examines Pittsburgh's history and rapid growth running the last 150 years. Use the slider to move from year to year, or select an individual year.

David Rumsey Historical Map Collection

A website where you can explore the extraordinary Historical Map Collection of David Rumsey. Experience over 67,000 historical maps in his collection.

Landsat is a time machine
First satellites to provide continuous global observation

Landsat sees Earth in a unique way. It takes images of every location in the world to reveal Earth's secrets, from deforestation patterns, to agricultural trends, to volcanic activity, to urban sprawl. The Landsat program started collection with early sensors in the 1970s and continues with the current Landsat 8 mission. Since every part of the earth is captured every couple of weeks, this enables us to see and analyze how places change over time.

Publicly shared imagery
The USGS manages the Landsat data program and makes the imagery freely available for everyone. This collection has been continuously updated with new scenes from various Landsat sensors for over four decades, resulting in an amazing, historical earth imagery resource.

New Landsat scenes are being collected daily. As new scenes are generated, they are added to a dynamically growing image mosaic containing millions of existing Landsat scenes in the shared database, providing extraordinarily useful information for historical comparisons.

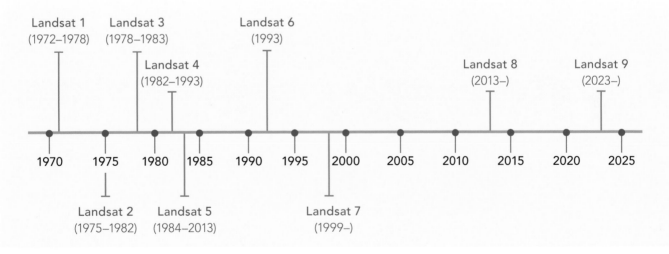

Map, measure, and monitor everything
Since the start of the Landsat program, a number of government agencies worldwide have launched their own missions—MODIS, the European Space Agency's (ESA) Copernicus program and its recent Sentinel-2 satellite pair, and many more—to continuously gather imagery and publicly share earth observations. New missions are being launched regularly, providing a growing collection of time series earth observations from space: a macroscope for the planet.

Historical map mosaics

Collections are easy to compile

Many maps are singular in nature, while others belong to a large map series or map collection. The USGS topographic map series, flood map series, insurance map series, aerial photos from specific missions of the past, the David Rumsey collection, the storehouse of historic maps from *National Geographic*—all are examples of map collections that can be used to enrich your GIS.

A useful approach for organizing and providing access to large historical map collections is to generate an image mosaic of your collection. The properties of each map—its name, creation date, spatial reference, and other characteristics—are recorded as attributes and used to create a seamless mosaic dataset. Mosaics help to bring your entire map collection to life in your GIS, enabling numerous applications and uses.

USGS Historical Topographic Map Viewer

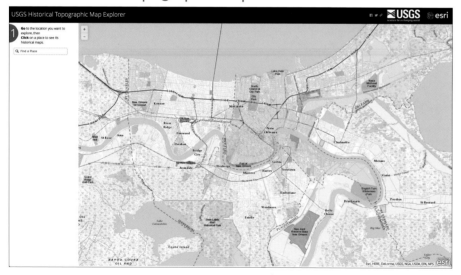

The USGS Historical Topographic Map Collection includes all scales and all editions of the more than 175,000 topographic maps published by the USGS since its inception in 1882.

Use the slider to change the transparency of various historic maps. Right-click a map to download and share it with your friends and colleagues.

Search or navigate to your area of interest and click the map to see a timeline of the maps that overlap it. Then click any map in the timeline to add it to the display and begin exploring.

Two hundred years of Dutch topography
Created and shared by the Dutch Cadastre

The Dutch Cadastre (The Netherlands' national mapping agency) generated a series of precached maps at numerous map scales (and for each year) for the entire country, stitching together the comprehensive collection of historic topographic maps. It is a magnificent cartographic treasure for the Dutch people and Dutchophiles everywhere.

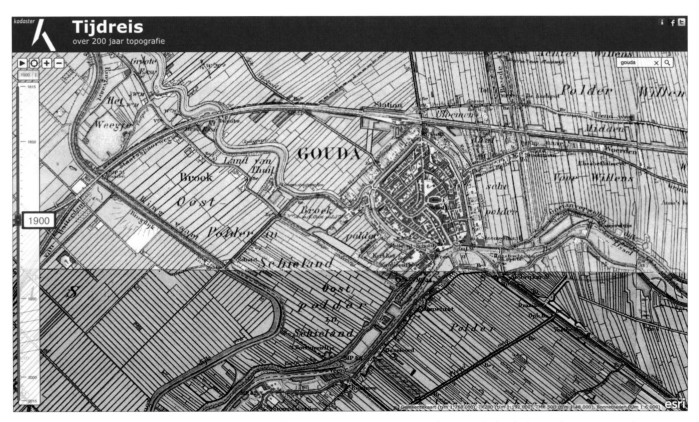

The Dutch have a history of cartographic excellence. This innovative application linked above brings together 200 years of maps into a single interface that makes comparing them as simple as sliding the time control on the left side. Use the thumbwheel on your mouse to zoom in and out.

Event-driven analysis
At the right time and the right place

GIS users seek to understand the results of an event by comparing the most recent situation to a previous state that is days or hours in the past. Common applications of near real-time imagery include coordinating emergency response, performing damage assessments, monitoring forests and agriculture, conducting military operations, and much more.

The term "near real-time" refers to time ranging between just before an activity or event and up to two to three days after the activity or event has occurred. This time range is often described as either real-time or near real-time, depending on how close or far (in temporal terms) you are from the event around which people are responding.

When considering event-driven imagery, it is important to determine how frequently you need to observe conditions affecting the event. During a hurricane, frequent images of the area can help you detect subtle changes in the storm direction and speed. In other words, the ability to be looking at a situation in a specific location can significantly aid the effort to defend life and property. GIS analysts assist in the fight by sampling such changes at a higher frequency, since timely decisions are likely to be more effective.

Disaster preparation and response

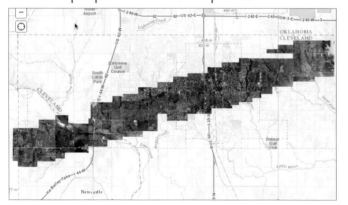

Ultimately this near real-time imagery is distilled into actionable information within a GIS. In this example of the tornado that struck Moore, Oklahoma, in 2013, imagery was flown at six-inch resolution within a day of the event and published online for responders and citizens alike.

Forecasting areas susceptible to wildfire

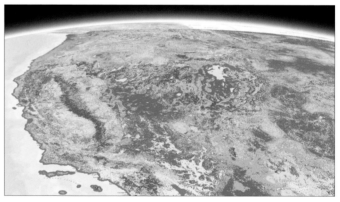

This map was developed by the US Forest Service and Fire Modeling Institute to help inform assessments of wildfire risk or prioritization of fuels management needs across large landscapes. The map depicts the relative potential for those wildfires that would be difficult to contain.

Weather forecasting in GIS

When meteorologists look at satellite imagery, they see more than just current weather. They peer into future weather as well. Meteorologists can analyze current conditions and forecast what will happen next.

Snow forecast map

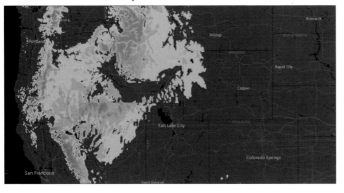

This map shows the latest forecast for snow accumulation from NOAA for the United States. It is designed to answer the question, When, where, and how much snow will accumulate in the next two days?

US precipitation forecast

This map displays predicted precipitation for the next 72 hours. Data is updated hourly from the National Digital Forecast Database produced by the National Weather Service. The dataset includes incremental and cumulative precipitation data in six-hour intervals.

Documenting effects of extreme weather

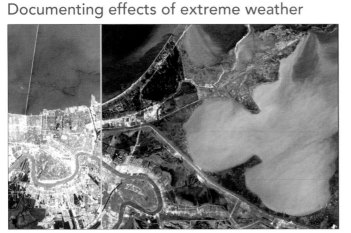

Hurricane Katrina, the most destructive storm in US history, spawned a decade of wrenching disruption throughout the region. Eighty percent of New Orleans was flooded. Click the Flooding tab to view before-and-after scenes in New Orleans.

Before and after the 2015 tornadoes in Illinois

Use the swipe tool in this story map to view the destruction caused by the April 9, 2015, multiple-tornado event in Fairdale, Illinois.

Imagery helps to see into the future

For future scenarios like climate change or sea level rise, effective response begins in the present—well in advance of the potential results and impacts that can occur. GIS makes it possible to run complex models into the distant future, enabling us to better understand the potential impacts.

Climate and weather forecasting

The landscape system layers that comprise key natural resources information in the Living Atlas include a range of climate prediction maps. You can access these by visiting the Climate & Weather gallery online.

Exploring our dynamic world

There can be no argument that the planet is changing rapidly. People born in the twenty-first century will experience more dynamic change in a shorter period of time than that experienced by at least the last few dozen generations. Imagery and GIS help people understand and share what these impacts could be and how we might mitigate the situation.

Atlas for a Changing Planet

Our planet is dynamic. This story map provides a way to investigate and better understand many of the dynamic facets of our ever-changing world. It illustrates the analytical and interpretive power of applied imagery as it is integrated with other geographic information for illustrating planetary change.

The King Tides Project: See the Future

King tides are the highest high tides of the year, occurring when the sun and moon are in alignment and closer to Earth, creating greater gravitational pull on our ocean's waters. Citizen scientists capture images to help determine what future sea levels will be and what is at risk from sea level rise and the impact of King tides.

Connecticut's Changing Landscape

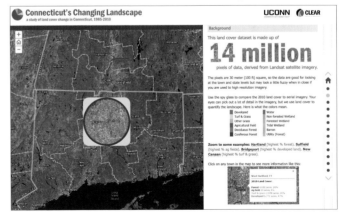

The University of Connecticut uses remotely sensed imagery to quantify and understand land-use change over a 25-year period from 1985 to 2010.

Forecasting the future

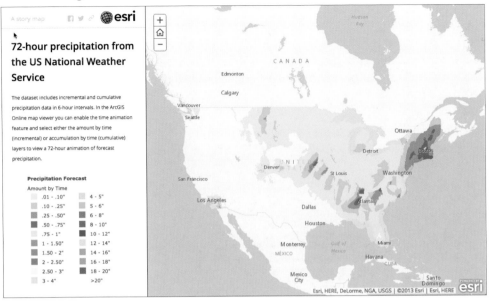

The dataset behind this map includes incremental and cumulative precipitation data in six-hour intervals. In the ArcGIS Online map, you can enable the time animation feature and select either the amount by time (incremental) or the accumulation by time (cumulative) layers to view a 72-hour animation of forecast precipitation.

Planning for future generations

GIS makes it possible to run complex time series models that simulate future conditions, providing the ability to envision future trends such as those for climate. This map forecasts the change in average annual temperature by 2050.

Case study: NOAA sea level rise

Our planet's future sea level depends on decisions that have yet to be made and cannot be predicted now. However, what's going to happen over the next 100 years is pretty well understood: Sea levels will rise by at least another three feet, even if we stop emitting carbon tomorrow, and by as much as six feet if we continue increasing emissions at our current rate. To help engineers, city managers, and concerned citizens understand what this means for their neighborhoods, NOAA's Office for Coastal Management has developed a Sea Level Rise Impact Viewer with imagery forecasting what the nation's coasts will look like as the ocean rises.

Fulton Street in Manhattan after six feet of sea level rise.

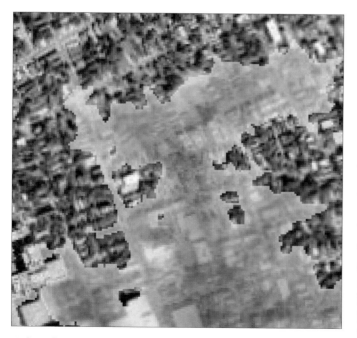

A four-foot rise in sea level would devastate Manmouth Beach, New Jersey.

This data viewer provides coastal managers, scientists, and citizens with a preliminary concept and understanding of sea level rise and coastal flooding impacts. The viewer is a screening-level tool that uses nationally consistent datasets and analyses. You can interact with the data and maps at several scales to help gauge trends and begin to articulate potential responses for different scenarios. A slider allows you to envision what various locations might look like with one foot, two feet, or up to six feet of sea level rise.

Tabs switch the map from natural color imagery to layers showing land-use classification and socioeconomic vulnerability, providing important context beyond just the visual impact of seeing buildings underwater. The application even forecasts how the land cover will change as sea levels rise, focusing on the impact to wetlands and marshes.

Quickstart

Get inspired by these examples of real-world problems explored through temporal imagery

▶ **Explore climate change projections for European cities**

Climate change projections suggest that European summer heat waves will become more frequent and severe during this century, continuing the trend of past decades. The most severe impacts arise from multiday heat waves associated with warm nighttime temperatures and high relative humidity.

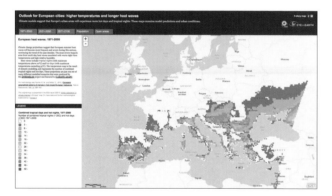

▶ **Explore aerial images before and after the April 2015 Katmandu earthquake**

On April 25, a severe earthquake struck Nepal. Satellite imagery of central Katmandu, captured two days after the quake, shows heavy damage in Nepal's capital.

▶ **See how maps and images come together to tell the story of the death of Abraham Lincoln**

April 2015 marked the 150th anniversary of the assassination of Abraham Lincoln. This map tour recounts the actions of John Wilkes Booth and his coconspirators through historical maps and imagery.

Learn ArcGIS Lesson

Depict land-use change with time-enabled apps

Over the last 40 years, Thailand has experienced tremendous shifts in land use due to a booming aquaculture industry. Widespread flooding of land to create shrimp farms has impacted sensitive ecosystems across the country, but particularly along the coast. As a GIS analyst for a nonprofit organization focusing on conservation and sustainable land practices, your goal is to find historical Landsat imagery for the Samut Songkhram province south of Bangkok in order to create a visual report of how the environment has changed over time. Your presentation will be shown to donors and investors to procure funding and promote the restoration of coastal ecosystems.

In this lesson, you'll build a presentation for identifying which region of Samut Songkhram province should be the focus of conservation efforts. You'll retrieve one image per decade since the 1970s from the Living Atlas Landsat archive for the entire study area. Once you have the images, you will analyze the available multispectral data to enhance vegetation, land, and water. Then, you will configure the time-animation tool in ArcGIS Online, and create custom time-aware apps for publishing your observations.

Start Lesson

Esri.com/imagerybook/Chapter7_Lesson

Build skills in these areas:

▸ Adding Landsat imagery to a map

▸ Enabling and configuring time animation

▸ Filtering historical satellite imagery

▸ Changing band combinations

▸ Creating and sharing time-enabled web apps

What you need:

▸ Publisher or administrator role in an ArcGIS Online organization

▸ Estimated time: 40 minutes

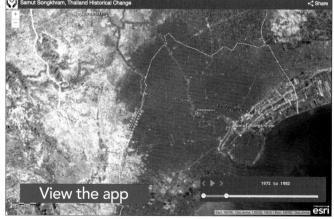

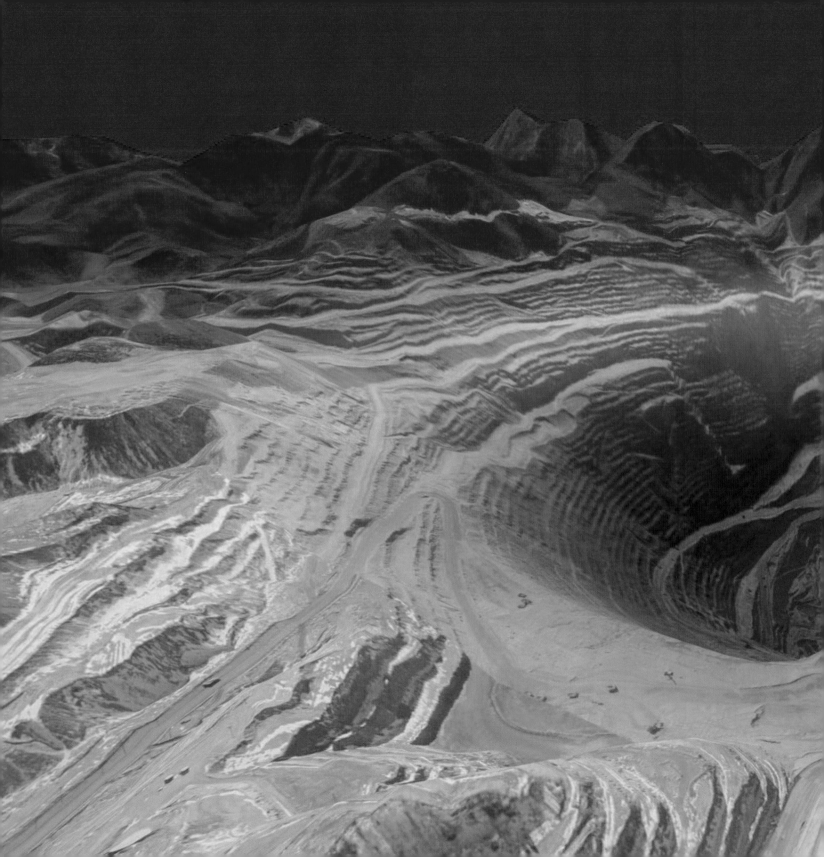

08

The Ultimate Big Data
Managing imagery information is a big data challenge

Imagery is big data and has always presented a challenge to GIS users. In the early years, various approaches were applied to overcome serious limitations in computing and data handling. These were the result of the enormous data volumes and the pace at which imagery could be collected. Only recently with the advent of new computing architectures in the cloud, along with advances in GIS software, have systems been able to leverage the massive opportunities provided by the volumes of imagery.

Imagery is big data
It's about geographic and temporal breadth and depth

Virtually 80 percent of the world's data has the capacity to be mapped as data layers in your GIS. This includes imagery and remote sensing information, an enormous array of feature layers, descriptive and tabular databases, full motion video, real-time and historical collections, sensor feeds, and massive point clouds of observations. What's surprising to learn is that the vast bulk of this data is sourced from imagery and remote sensing platforms (such as data from Landsat shown below). With the recent trend in web and cloud computing, the world's GIS information is becoming more broadly accessible to increasing numbers of people.

We are all tapping into a common "GIS of the world."

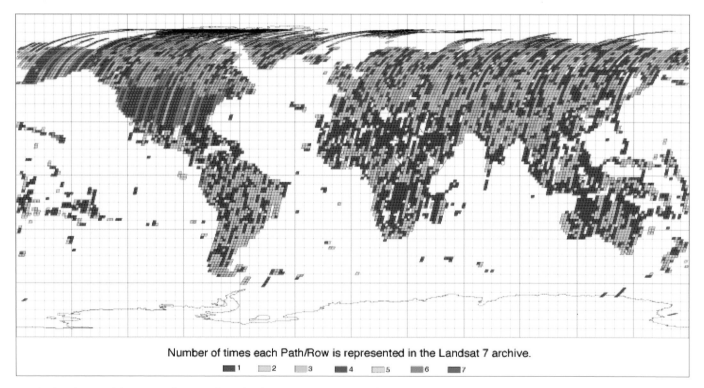

Number of times each Path/Row is represented in the Landsat 7 archive.

■ 1   □ 2   □ 3   ■ 4   □ 5   ■ 6   ■ 7

Each day the world's constellation of earth observation (EO) satellites maps, monitors, and measures our planet, generating enormous volumes of data. This map illustrates the relative volume of imagery archived during the first 112 days of the Landsat 7 mission. Although the mission was still young, certain trends were emerging. The United States (including Alaska) is quite green because every imaging opportunity was being exploited. North Africa is mostly desert and appears red because of low imagery demand. Northern Asia is mostly red and yellow due to recorder constraints in the Landsat 7 acquisition calendar.

Planet probes
Massive information gathering from the sky

For decades the earth has been monitored and observed from the sky, increasingly measured, sensed, and photographed by thousands of sensors mounted on satellites, aircraft, and drones. And the pace of these collection activities is expanding. NASA alone operates 19 different earth observing missions (10 of which are depicted here.)

There is a fire hose of imagery coming from this technological constellation. Petabytes of observations and scientific measurements are being collected, downloaded, and put to work in the form of imagery in GIS systems worldwide.

And it's growing exponentially.

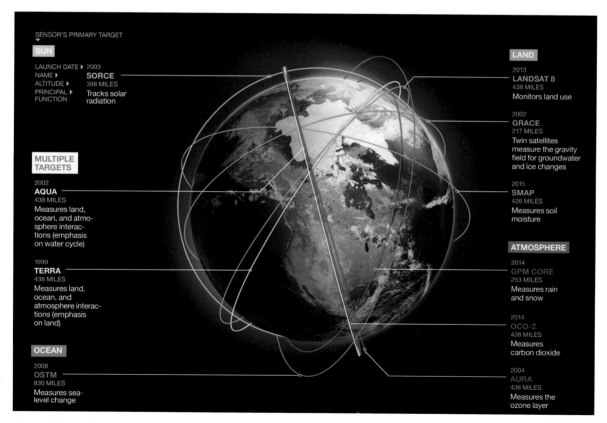

SENSOR'S PRIMARY TARGET

SUN

LAUNCH DATE ▶	2003
NAME ▶	**SORCE**
ALTITUDE ▶	398 MILES
PRINCIPAL ▶ FUNCTION	Tracks solar radiation

MULTIPLE TARGETS

2002
AQUA
438 MILES
Measures land, ocean, and atmosphere interactions (emphasis on water cycle)

1999
TERRA
438 MILES
Measures land, ocean, and atmosphere interactions (emphasis on land)

OCEAN

2008
OSTM
830 MILES
Measures sea-level change

LAND

2013
LANDSAT 8
438 MILES
Monitors land use

2002
GRACE
217 MILES
Twin satellites measure the gravity field for groundwater and ice changes

2015
SMAP
426 MILES
Measures soil moisture

ATMOSPHERE

2014
GPM CORE
253 MILES
Measures rain and snow

2014
OCO-2
438 MILES
Measures carbon dioxide

2004
AURA
438 MILES
Measures the ozone layer

Every day, voluminous imagery and remote sensing information is being collected and put to work in hundreds of thousands of GIS systems worldwide.

The cloud changes everything
Cloud computing enables on-demand imagery

Recently, the computing landscape has gone through a radical transformation, led by the emergence and rapid adoption of cloud computing and the smartphone app revolution. The complexity and investment constraints required to install and implement high-capacity servers within your organization to manage imagery and its big data have been removed. With cloud computing, you can now turn a switch and get a new server in the cloud onto which you can deploy your processing as needed.

One of the big data breakthroughs is how ArcGIS directly connects to and interoperates with an authoritative, collective GIS imagery repository of immense proportions. The new web GIS paradigm enables you to work with your geographic information, in concert with everyone else's geographic layers, using the online, cloud-based ArcGIS platform. Cloud computing enables massive storage and flexible computing that can expand to meet your computational needs. It provides tools for big data use and integration on demand.

Over the past four decades, Landsat has amassed over a petabyte of data, with over four million scenes and counting. It takes images of every terrestrial location on earth approximately every 16 days. Recently deployed onto the web, the Landsat archive is only a click away. In this app, use the icons on the left panel to switch among different spectral views. Some are combinations of bands and others are commonly computed indices. ArcGIS automatically enhances the image based on your current view, so try zooming and panning. You can also use the time tool (the clock icon) to explore the available time periods of your own area of interest.

Case study: AWS Landsat
An ambitious program to share Landsat data

In 2015, Amazon changed the game in satellite imaging when it announced the availability of Landsat imagery on Amazon Web Services (AWS), its popular cloud computing platform. Under this program, Amazon will host one petabyte of Landsat imagery.

All Landsat 8 scenes from 2015 and 2016 are available, along with a selection of cloud-free scenes from 2013 and 2014. All new Landsat 8 scenes are made available each day, often within hours of capture. AWS has made Landsat 8 data freely available on Amazon S3, and served via ArcGIS, so that anyone can use the on-demand computing resources of ArcGIS to perform analysis and create new products without needing to worry about the cost of storing Landsat data locally or the time required to download it.

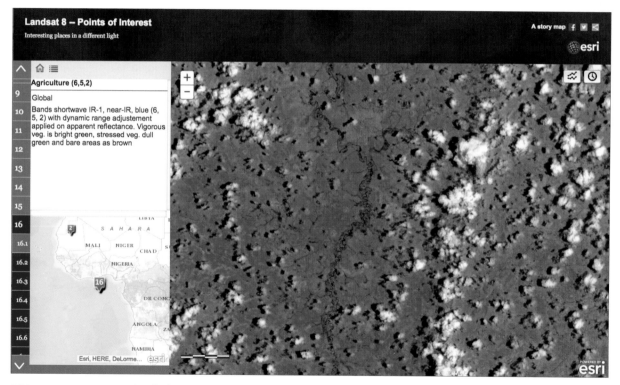

This story map provides links to interesting locations using different band combinations of Landsat 8 imagery. For each location, a general description is provided. The maps are live, and you can navigate to any location in the world and see the same bands. These maps also include a couple of widgets you can use to select particular scenes from the overlapping imagery and explore the world.

The early days of imagery
A short history of GIS and image processing

Imagery has faced computing challenges since its inception because of the immense data volumes and the continuous feed of sensor observations from various satellites and aircraft. In the early years, and up until very recently, most users did not have access to enough computing power, or nearly enough storage capacity, for all of the big data. In addition, only a few enterprise systems, with massive capacities, were strong enough to meet the storage and processing needs for exploiting and gaining the greatest benefits from working with imagery.

For practical reasons, most GIS specialists and imagery analysts developed various procedures to maneuver around the computing issues. Many of these continue to this day.

Image processing and GIS were segregated
In spite of the fact that rasters are a universal GIS data type that enables powerful modeling and spatial analysis, it was difficult to integrate image processing systems and GIS due to big data challenges. In most system implementations, image processing systems and GIS were segregated.

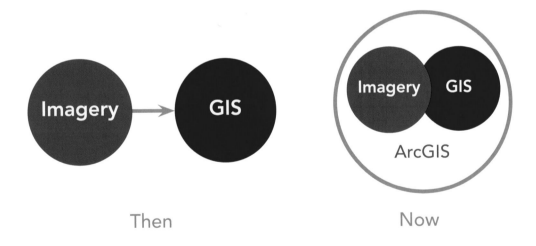

Then   Now

Historically GIS and imagery were separate, but they have since become integrated in ArcGIS.

Image processing was performed in one system, and then the results would be fed into the separate GIS. In many cases, image processing and preparation were implemented by separate contractors outside of the GIS organization and then returned. The desired capability to work with and integrate multiple layers was more challenging than practical.

Sharing most image layers required compression with lossy formats

A number of creative techniques were used to integrate immense amounts of imagery information. One commonly used approach involved compressing the image data and precomputing the results—for example, using MrSID or JPEG 2000 compression. This enabled GIS users to readily deploy a coverage of imagery for their areas of interest. In fact, many users still utilize MrSID and JPEG 2000 today to share focused imagery collections to remote crews.

When you convert an image file to a lossy format such as MrSID or JPEG 2000, you're throwing away some of the data that is not needed for image display. You get a new file that is much smaller in size and more easily shared. However, the lost data may be crucial for some image analysis workflows.

The problem with pre-processing and lossy compression, is that the original data is required for most analytical operations. Another is that pre-processing and compression is disruptive and users are then unable to keep up with the continuous data feeds.

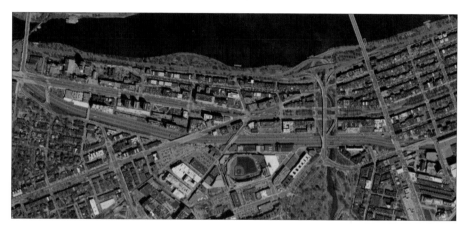

True color imagery for the area around Fenway Park in Boston, Massachusetts, at 0.3-meter resolution provided as file-sharing downloads by USGS and MassGIS. This was created using JPEG 2000 at a 16:1 compression.

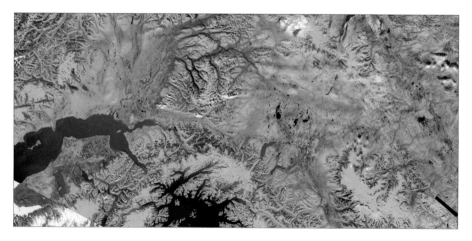

This Global Land Cover mosaic delivered using MrSID compression is used for visualizing land-cover change and research using remotely sensed satellite data and products. The MrSID compression reduces the image data sizes by as much as 20:1. However, this also leaves behind the original image bands, removing many analytic capabilities.

ArcGIS supports big imagery on demand
The culmination of an image processing revolution

Earth imagery presents a number of computing challenges, including how to store and process the vast and dynamically growing numbers of individual scenes over space and time (the proverbial summary of the problem: "four million scenes and counting"). Here is a short review of some of the key capabilities that enable ArcGIS to support and make the most of this "big imagery."

High-capacity image storage with mosaic datasets
Mosaics datasets allow you to store, manage, use, and share small-to-vast collections of imagery and raster data. A mosaic dataset is a flexible and scalable way to assemble and use a collection of original imagery source files ranging in size from quite small collections to massive amalgamations. For example, a mosaic dataset might represent a small collection of a few dozen orthophotos from a personal drone mission you just flew yesterday. Or at the other extreme, it might include a complete, continuous image collection of the world—such as the entire Landsat 8 collection stored and made available in Amazon Web Services or the Copernicus Sentinel-2 imagery of the world. Hundreds of new scenes are being added to both of these collections daily. Mosaics enable flexible storage, management, and use of these scenes as cohesive image collections.

Mosaic datasets in ArcGIS are versatile and elastic and can be updated at any time. They have the capacity to process and manage millions of imagery files as a single integrated dataset while ingesting ongoing updates and additions as satellites continually collect additional scenes. Mosaic datasets are always available and ready—an at-your-service data source. This is important given the growing data volumes coming from the ever-expanding series of earth observation platforms. Imagery from all satellite systems can be managed and served using mosaic datasets.

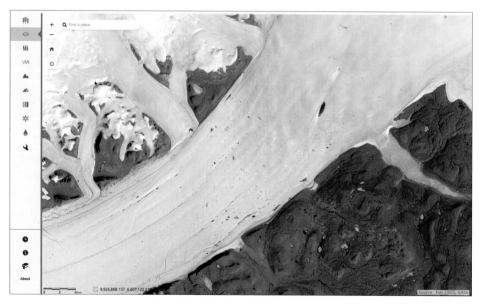

Mosaics handle fast-growing earth observation datasets, like the Landsat Arctic collection. As new scenes are added daily, the collection continuously grows in size and scope. Online, you can zoom to an area of interest. For example, open the Time Line tool, and click along the time line to explore the full collection.

Distributed raster data storage

ArcGIS is able to work with multiple scalable data storage types. You can work with your imagery data where it currently resides on your existing computers and servers within your organization. Alternatively, you can use distributed data storage in the cloud—available, for example, with Amazon Web Services (AWS) or Microsoft Azure. These storage options provide near-limitless storage scalability.

Distributed, scalable raster analytics in ArcGIS

Distributed raster analytics in ArcGIS offer flexible, high-capacity image computing. Raster analytics in ArcGIS are designed to scale with your organization's demands. Spatial analysis models and image processing computations are applied where your data is stored—on your local networks, in your servers on your enterprise networks, or when you leverage cloud storage to further improve the scalability of your distributed image analytics. Your analytic results are written in parallel to the distributed data stores.

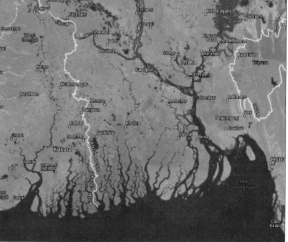

Distributed raster analytics provide the flexibility to scale up to get the job done on demand when resources are needed. Calculating a multiscale Global Vegetation Index from 7,422 Landsat Global Land Survey (GLS) scenes presents a massive computional task that was never before possible on regular personal computers. Now, using processing on multiple virtual machines from AWS, this multistep big data processing sequence can be completed in under three hours (44 scenes per minute). In the map, high vegetation density is shown in green. Oceans and other water bodies show red, indicating a total absence of vegetation.

Big imagery is available to all users

ArcGIS distributed raster analytics enable you to optimize your big imagery work when you need it. As the size of your imagery data increases, a single machine may not be sufficient to store the data or provide enough computational throughput. ArcGIS provides the flexibility to scale out your imagery work when you need it.

Distributed data stores and analytics in ArcGIS are optimized for cloud-based cluster computing. This enables parallelization of massive image computing, leveraging the inherent scalability of mosaic datasets. Large processes can be spread over multiple machines all running in parallel to significantly reduce processing times.

Big imagery enables big understanding

Cloud computing and ArcGIS let you dynamically expand your image applications. For many GIS practitioners, a unique capability of GIS is how it unlocks imagery analysis and exploitation—delivering insight through computation and modeling. That's the power of GIS. These might be applied on hyperlocal projects—like your drone mission for your agricultural field. Or they might span processing across an entire satellite collection or sophisticated time series simulation modeling of climate forecasts.

Distributed analytic computing scales across many scenarios.

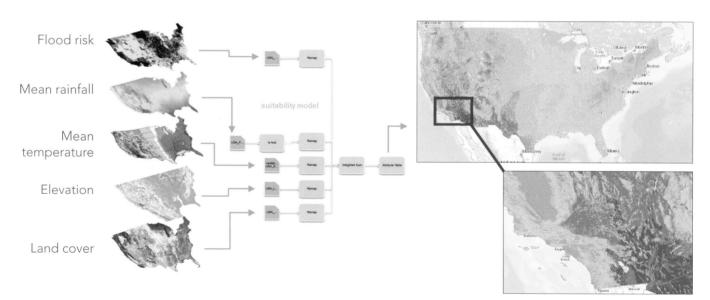

Distributed raster computing is scalable. You do not require access to massive computing resources to get dramatic results. Even modest server configurations provide compelling performance that continues to improve. Most important is flexibility—every ArcGIS user can scale computing to meet their needs.

Thought Leader: Kathryn Sullivan

Twenty-four years ago, I worked with emergency managers after Hurricane Andrew obliterated huge portions of Florida's infrastructure. We showed up with large paper maps and hand-edited data for days, locating telephone, water, sewage, and electricity nodes for responders desperate for the full picture. We basically made a rudimentary GIS by hand. It's amazing that web-based platforms now let us synthesize and visualize data from dozens of data sources in real time. Such GIS tools enable smarter, faster, more informed decisions for highly diverse needs.

Today we depend on a soaring volume of data to underpin the environmental intelligence vital to protecting lives and livelihoods and fostering economic resilience. But while data drives environmental intelligence, that's just the start. The challenge is to connect that data to practical needs, to make it actionable from Wall Street to Main Street, and for heads of households and heads of state. The data must reach those who need it, when they need it, and at the right scales and resolution. The data must be discoverable, searchable, and easily retrievable. In short, it must work for you and for everyone.

When it does, the payoff is enormous. With Esri tools and NOAA data, the reinsurance community can verify that claims are in sync with accurate severe weather assessments. Esri story maps make practical sense of storm predictions and intensity, sea level rise, fisheries impacts, and much more. NOAA is expanding America's river-forecasting capabilities 700-fold, to about 2.7 million more locations. Scientific data enables all of these environmental intelligence products. They provide the foresight to hedge risks. And they yield an understanding of our planet that can help shape the way we live on it.

Kathryn D. Sullivan, NOAA Administrator and Under Secretary of Commerce for Oceans and Atmosphere, and former NASA astronaut.

Yet great volumes of federal data key to environmental intelligence remain untapped. NOAA observations alone provide some 20 terabytes every day—twice the data of the Library of Congress' entire print collection. But just a small percentage is easily accessible to the public. As America's environmental intelligence agency, NOAA is committed to getting much more of our data out the door. We are eager to put data from labs and hard drives into the hands of those just as eager to transform it into products, services, and predictions that can add significant value across every sector. There is huge potential environmental and economic value in this data.

NOAA is partnering with Amazon, Microsoft, IBM, Google, and the Open Commons Consortium to tap that potential. Esri is an active player as well. We can imagine a private weather enterprise that, by combining previously separate data streams, introduces a vulnerability model and grows in value from $2 billion to $20 billion. Think of the tremendous support transforming millions of data points from satellites into long-range rain forecasts would give to farmers and businesses struggling with drought.

Just 20 years ago, we were piecing together data points by hand. Five years ago, 90 percent of today's data was yet to be generated. Now we're innovating in the cloud, experiencing Earth with a wider lens and in fresh new ways. Dynamic opportunities are on the horizon. We hope you'll help to shape them.

Watch video of Sullivan bringing science to life

Gallery of big imagery

This gallery illustrates an array of the various kinds of imagery and remote sensing information that is being collected daily. It provides a mind-boggling illustration of the opportunities for putting imagery to work in your GIS.

Sentinel-2

Sentinel-2 Bands	Central Wavelength (µm)	Resolution (m)
Band 1 - Coastal aerosol	0.443	60
Band 2 - Blue	0.490	10
Band 3 - Green	0.560	10
Band 4 - Red	0.665	10
Band 5 - Vegetation Red Edge	0.705	20
Band 6 - Vegetation Red Edge	0.740	20
Band 7 - Vegetation Red Edge	0.783	20
Band 8 - NIR	0.842	10
Band 8A - Vegetation Red Edge	0.865	20
Band 9 - Water vapour	0.945	60
Band 10 - SWIR - Cirrus	1.375	60
Band 11 - SWIR	1.610	20
Band 12 - SWIR	2.190	20

The Sentinel-2 mission by the European Space Agency (ESA), the European Commission, and industry comprises two identical satellites positioned at 180 degrees to each other. Sentinel-2 performs terrestrial observations for applications such as agriculture, land-cover mapping and change detection, forestry, and environmental monitoring. The multispectral data has 13 bands, and spatial resolution of these bands varies between 10, 20, and 60 meters as shown in the table at right. Each of the twin satellites has a revisit time of 10 days, so in effect every location on earth is revisited every five days.

Oceans

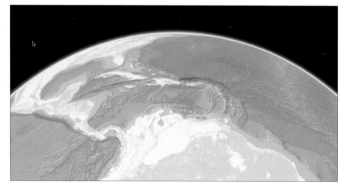

Detailed Earth mapping hasn't stopped at land's end. With a steady increase in the number of ocean-measuring missions, a fine-grained subsurface map is emerging.

Drones

Drones fly low to the ground and collect large payloads of high-resolution imagery. Managing and storing all this data is a challenge answered by Drone2Map.

Lidar

Raw data | Color classified | Points assigned RGB values

Lidar data involves a unique big data challenge because of the sheer number of points in even a small scene, which is then multiplied by the typically large study areas. All of this is compounded by the need for multiple lidar producrs, each with its own characteristics and applications.

Full motion video

Whether captured by UAVs or aircraft, video is an important data source that provides a dynamic view of unfolding events and the state of features on the ground. Full motion video (FMV) capabilities in ArcGIS let you view, manage, analyze, and share video content, including live feeds.

Historical air photo collections

Statewide aerial photographs were first acquired for Illinois from 1937 through 1947. This collection contains approximately 33,500 photographic paper prints that were scanned and converted to digital imagery.

Quickstart

Mosaic dataset layers

Managing imagery in ArcGIS Pro starts with the mosaic dataset. It is a catalog of data that allows you to manage, analyze, and map all of the scenes from your collection.

Mosaic datasets reside within a geodatabase and consist of three layers: boundary, footprint, and image.

- The boundary layer shows the extent of all the raster datasets within the mosaic dataset as a single polygon, or as a multipart polygon if your collection of imagery is not contiguous.

- The footprint layer shows the extent of each individual item in the mosaic dataset as a distinct polygon. The footprint attribute table is the catalog of all the images in the mosaic dataset in addition to any associated overviews. Within this table, you will be able to sort your imagery based on any of the attributes, such as cloud cover, acquisition date, or any of the sensor characteristics.

- The image layer controls the management, analysis, and map display of the mosaic as one integrated raster layer. Display and rendering properties, such as stretch, band combination, resampling, and mosaic method, are applied to the mosaic.

Mosaic datasets are capable of dynamic and on-the-fly processing. Because the imagery is processed as it is accessed, you can create multiple products on the fly from a single source. You can take imagery from multiple sources at multiple resolutions and create a virtually seamless mosaic. All of the individual rasters that together make up the mosaic dataset can be accessed through a database, allowing you to determine how to display overlapping rasters.

Imagery and raster in ArcGIS Pro

Manage a collection of imagery and raster datasets

Learn ArcGIS Lesson

Download imagery from an online database

In this lesson, you'll assume the role of an urban planner looking for imagery of Singapore, a massive metropolis bounded by the confines of a small island. Singapore's high population density (nearly 8,000 people per square kilometer) necessitates tight control over the city's development. Using the LandsatLook app, you'll search the USGS databases for a relatively recent image with minimal cloud cover. You'll download the image and add it to a map in ArcGIS Pro. You'll then change the default band combination to display the image more clearly.

▸ Overview

The Landsat satellite program, run by the US Geological Survey (USGS) and National Aeronautics and Space Administration (NASA), has continually collected imagery of our planet's surface since 1972. This imagery is freely available for download from the USGS website. However, with over four million Landsat images to choose from, finding the best one for your needs can be difficult.

▸ Build skills in these areas:

- Finding and downloading Landsat data
- Displaying Landsat data in ArcGIS Pro

▸ What you need:

- ArcGIS Pro
- 900 MB of hard drive space
- Estimated time: 1 hour

Start Lesson

Esri.com/imagerybook/Chapter8_Lesson

The use of Landsat data has been historically limited to the scientific and technical user communities. The LandsatLook Natural Color image product option was created to provide Landsat imagery in a simple, user-friendly, and viewer-ready format, based on specific bands that have been selected and arranged to simulate natural color. This type of product allows easy visualization of the archived Landsat image without any need for specialized software or technical expertise.

09

The Future Is Now

The map of the future is an intelligent image

If you've read this far, you are fully aware of the massive transformation taking place in GIS and imagery. Never before have so many information feeds of imagery and geographic information been accessible to the rapidly growing GIS community—with no computing or storage limits. This is true not just for extremely large enterprise computing configurations in big organizations, but also for virtually everyone via ArcGIS in the cloud. You are limited only by your imagination of what might be possible.

We are entering the human era of GIS

With the growth in cloud computing, along with advances in ArcGIS and the explosion of imagery and remote sensing feeds, it's an exciting time to be a GIS user. A great transformation is underway. GIS is becoming more social, more accessible, and more effective. Every day, we are seeing how GIS can make a real difference and add meaning to people's lives. GIS is becoming a human activity—how people apply ArcGIS as a framework to collaborate with each other, to work more efficiently, to make better decisions, to increase understanding, to communicate better, and to make a real difference in the world—to intelligently effect change. And imagery is playing a big part in this transformation.

What is clear is that these new environments invite and encourage—and, some would say, demand—collaboration and information sharing across organizations and communities. We believe that these opportunities are so significant that the concept and vision for GIS has forever changed. Not only are you working on your GIS for your organization, but your work is also a vital part of the larger "GIS for the world."

Your own GIS is simply your view into this larger collaborative system. It's a two-way street. You consume information that you need from others, and in turn, you feed your information back into the larger ecosystem for exploitation and use by other organizations.

Can there be any doubt that imagery is leading the way for this revolution in information sharing and use?

Imagery within ArcGIS is driving explosive growth as GIS moves to the cloud with open data and community engagement. The US Department of Labor has projected the growth in GIS jobs over the coming decade to be among the top three in the computing and technology sectors.

Imagery is enabling the GIS for the world
Modern GIS is about participation, sharing, and collaboration

GIS has grown to encompass use by over 350,000 organizations worldwide and is being applied in virtually every field of human endeavor. Imagery provides the engine that lights up GIS, and it is enabling unprecedented adoption across the world. That's the power of imagery—we all understand something when we can see it.

Since the early days in GIS, people realized that to be successful they would need imagery and data from sources beyond their immediate workgroups. People quickly recognized the need for data sharing. Open GIS and data sharing were gaining traction quite rapidly across the GIS community, and these features continue to be a critical aspect in GIS implementation.

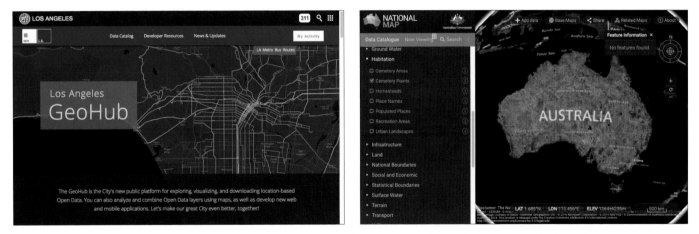

Many governments at all levels are opening up access to their geographic information, including their imagery collections, because they recognize the many benefits for their citizens.

GIS: The integration engine

GIS provides a kind of integration engine, and this is a profound idea. By all of us focusing and working on our own geographic information systems, we're assembling this very promising, comprehensive GIS for the world. This is growing and getting richer every day. There's a global collection of imagery information that's being assembled and updated daily. These feeds and services are further built out by the entire GIS user community—all in a series of information layers that reference onto the earth, making it simple to integrate information from multiple sources.

Imagery and GIS are web-accessible

GIS is moving to the cloud, to this great computing network that makes information available, enabling all of us to have access to these rich collections of information. Every layer has a URL—a web address—that is findable and makes the data easily usable. You now simply reference a data layer's URL to bring it into your GIS and use it for real work. Mash up layers from different sources, save it all into a web map, publish it as an app, and all of a sudden you've created a brand-new, powerful information product that can be shared with virtually anyone. This new era of geo-authorship enables us to take a point of view, make a case, and persuade and influence others.

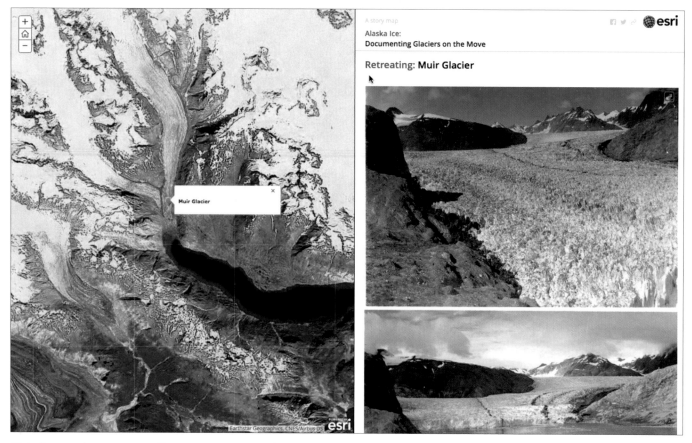

This focused story map offers convincing evidence about a warming Alaska. The range of shared imagery and GIS applications by many in the GIS community is profound. These resources enable thousands of organizations to leverage this content from the greater GIS community in their daily work. When you're applying ArcGIS, you'll soon begin to realize and appreciate the value of information sharing throughout the community and how much it enriches and enables your work.

Imagining the GIS of the future
The future is now

Imagine a future where every GIS user has access to a catalog of imagery, remote sensing information, and other geographic layers shared by many organizations for all areas of the world. Further imagine that this imagery catalog is being updated with new scenes every day. And that new sensors are being launched on a daily basis. And that much of this information is being publicly shared. The future is now.

This modern GIS that has emerged is participatory and social. Given our society's information needs, these modern systems don't just encourage participation—the system users demand it. It's worth recognizing and drawing your attention to many of these key advances that are changing our view and our vision for what is possible with the synergy of imagery and GIS.

No limits on data storage and computing

GIS and imagery require not only big data, but also big computing. And any level of computing is possible now in cloud systems that you can turn on when you need them. Today, you can run your high-end analytics near your data sources. Storage capacity is virtually limitless. More is available when you need it. Users rarely buy large computing configurations—instead they gain access using cloud computing.

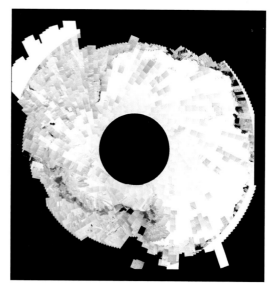

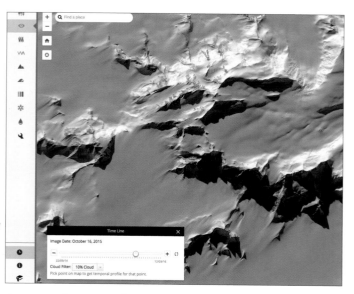

This image map depicts a small portion of the thousands of Landsat scenes gathered over time and organized into an image mosaic dataset for the Antarctic region.

A zoomed-in view of some of the available imagery for Antarctica. Note the time controller (toggled on with the clock icon) that allows you to click from one capture to the next. Never before have scientists had so much ready-to-use data at their fingertips.

Public image sharing

In the early years of image processing, many imagery sources were expensive to collect and would take months to gather. Today and in the future, many satellites and other platforms are delivering high-value information using multiband sensors, lidar, and many other data sources. Users now expect that highly suitable imagery will be available to support their work. Like never getting lost, thanks to GPS smartphones, or having cinema-quality large displays in our living rooms, 24-7 access to all the world's imagery is the new normal. The implications of this transformation are just starting to be recognized.

The city of Berlin, Germany, along with hundreds of other governments and organizations worldwide, has contributed its high-resolution, up-to-date orthoimagery to the ArcGIS Community Basemap.

Ready-to-use analytics

GIS is all about analysis and combining multiple layers using intelligent analytic models. ArcGIS provides a rich and varied collection of analytic tools for you to harness and apply to your work and problems. These analytics can be applied to data residing anywhere—on personal desktops, in large enterprise computers, or in ArcGIS Online in the cloud. Imagery analytics comprise one of the most dramatically expanding capabilities for big imagery processing.

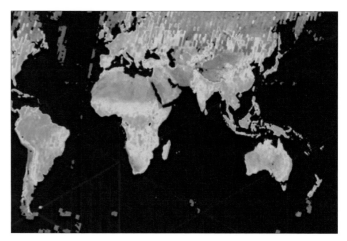

Imagery analytics are available everywhere. An emerging army of image analysts will compose and share models for all kinds of imagery computations. These will be made available in curated galleries for the diverse and growing GIS community.

Hyperlocal custom missions

The age of drones has arrived, along with our ability to fly personal missions to map our own small geographies and to apply GIS in whole new ways to a new range of problems—assessing street pavement; flying utility corridors; conserving fish and wildlife habitat; responding to emergencies and disasters; monitoring our climate, forests, and agricultural fields; and managing our small microgeographies.

Many users will fly their own local missions for use in their own private GIS applications. With Drone2Map, you can create an almost instantaneous GIS basemap foundation for your specific study area, on top of which you can begin to add and analyze all kinds of other GIS layers—"instant GIS."

Custom satellite missions

New sensors are being launched every day and are helping us build focused imagery collections. These contain spectral bands that enable analytic processing to divine information and results that enable us to better monitor and manage our operations and missions.

Shoe-box sized satellites being launched from the International Space Station. This fleet of nanosatellites— part of a cooperative mission between the US National Laboratory, NanoRacks, and Planet Labs—is intended to provide daily snapshots of the earth for a variety of applications and uses.

Converging technology trends

On-board analytics provide access to both raw and processed imagery

In the future, not only will you download raw imagery data from satellites, you will also download processed imagery in which analytic computations are performed on the fly to serve your imagery.

ArcGIS contains analytic capabilities that can be deployed virtually anywhere, from the sensor to your local computer to cloud processing of your intelligent imagery. Further, this analytic processing is happening in near real-time. So the information feeds you see coming from these platforms are more than raw image data feeds—they provide imagery intelligence that has already been processed into useful information to better support your operations and activities.

Computing limits will disappear

How many people do you think have access to 10,000-plus computers as Google does? And who could afford that? Actually, the reality is that all ArcGIS users are gaining access to massive computing resources in the cloud for distributed raster storage and analytics.

The emerging GIS of the world will provide a macroscope for Earth

GIS and imagery provide an integration engine by combining geographic information layers of many kinds. As community participation and sharing grows across thousands of organizations, the information and imagery layers in the GIS of the world become richer and more capable. ArcGIS, along with cloud computing, is enabling insight about our planet.

It's enabling users to integrate these layers both visually and analytically, providing new and deeper insights and perspectives to better understand and manage our world.

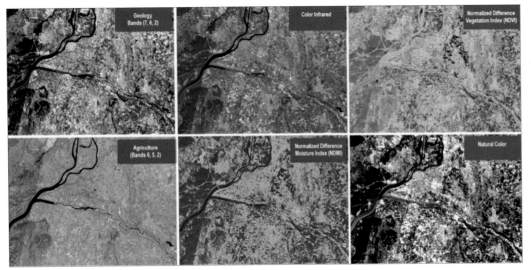

Multispectral imagery can be processed and represented for many purposes, each served directly from the sensor platform on the fly. Image processing algorithms can be applied to the raw imagery in order to represent the results you want to work with. In this example, Landsat 8 scenes are interpreted and simultaneously delivered as real-time image intelligence.

The future gets personal

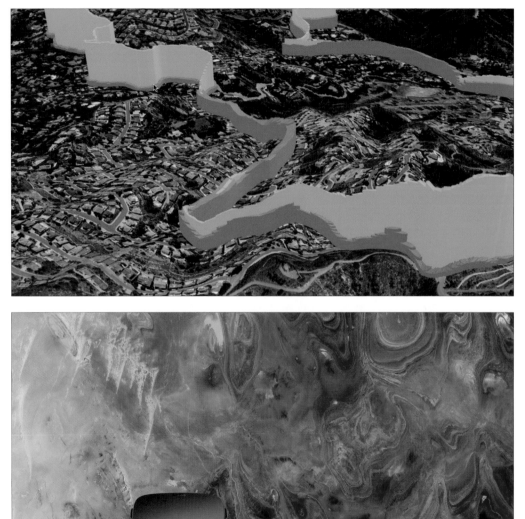

This navigable 3D web scene depicts an actual bicycle ride, with speed and heart rate information from a GPS device displayed as a 3D spatial graph, on a web scene of high-resolution imagery draped over elevation-exaggerated terrain.

Converging technologies are ushering in an era of nearly unimaginable immersive interaction with geoimagery. New applications for 3D headsets like Occulus Rift promise to redefine our understanding and very relationship with the world around us. Prepare to be amazed.

Creating GIS data from drones
2D gallery

The photos and video gathered from drone missions are captured either pointing straight down, for the purpose of creating orthocorrected mapping products, or oblique angles at buildings and other infrastructure, typically for inspection applications. Both approaches generate GIS-ready data. This page shows examples of the 2D GIS data outputs that are generated from Drone2Map sessions; the following page shows a gallery of 3D outputs.

Orthomosaic

For GIS users, an orthomosaic represents the ultimate ground truth. Orthomosaics offer the ability to validate existing map features and create new map features through manual or automated digitizing, and provide the GIS basemap for your study area in ultra-high resolution.

DSM

A DSM, seen here side by side with an orthomosaic of the same buildings near Lake Geneva, Switzerland, offers an unadorned glimpse of the real shape of things on the ground. It has a hillshade renderer applied.

Creating GIS data from drones

3D gallery

LAS data rendered in RGB

A LAS file is an industry-standard format for storing airborne lidar data. In this rendering, each point has been assigned an RGB value from the source photo, creating a photorealistic effect.

LAS data rendered as elevation

The same LAS point cloud can also be rendered based on attributes in the data, such as elevation as shown here. The red color is the highest elevation in the scene.

3D mesh

The 3D mesh is a composite of all the points, connected together in a constellation of triangles, or facets. This allows for a navigable 3D scene.

Oblique photo collection

The raw oblique photos taken by the camera serve multiple purposes. First, they are the source for 3D data layers. In addition, each photo also serves as a stand-alone inspection product.

Thought Leaders: Kathryn Keranen and Lyn Malone

GIS technology is changing the way our students learn

How many times have you heard a "digital immigrant"—someone born before social technologies came online—say something like, "My 5-year-old nephew had to show me how to use an app on my smartphone"? The majority of today's students have never experienced life without smartphones, desktop computers, tablets, laptops, or GPS devices. They have nearly always been digitally connected to the world through the Internet, mobile technology, and social media, which means they are the first true "digital natives"—a term used by the Pew Foundation that means they have never had to adapt to digital technologies. This has transformed how modern students interact with others, how they view the world, and how they use technology. Growing up digital has also significantly impacted how students learn.

GIS, global positioning systems (GPS), and remote sensing are fundamentally changing the way students learn about their world. The same tools that government agencies, businesses, environmental groups, and health care industries are using to solve problems and make decisions in their respective fields are now being employed by students to explore and address real-world issues in their classrooms.

GIS is a real-world, problem-solving tool. GIS users outside of the educational setting—those in government, business, natural resources management, health care, and so on—employ this powerful technology for that very reason. It is, by far, the most versatile and effective problem-solving tool available. Whether the subject is science, math, or economics; whether the classroom is elementary, middle school, or high school—when students learn with GIS, they learn how to use technology

effectively to holistically address questions and they are better prepared to meet the challenges of work and life in the twenty-first century.

Educational leaders have long argued for a problem-based approach to learning. Problem-based learning is open ended; it is a process of inquiry that examines alternatives and possibilities rather than memorizing a single, correct solution. The use of GIS in the classroom fosters problem-solving skills such as asking relevant questions; acquiring data with which to investigate those questions; observing and analyzing patterns in that data; drawing inferences and conclusions from data analysis; and acting on the findings of the investigation. GIS prepares students for the workplace, where problem-solving skills are essential. A GIS-enabled classroom is a stepping stone to the future. GIS, by its very nature, is interdisciplinary and serves as an ideal tool for the integration of knowledge and skills from varied core subjects. In GIS, mathematics and geography merge to form a foundation for the observation and analysis of data from history, civics, environmental science, biology, earth science, sociology, linguistics, literature, and other fields of study. Through interdisciplinary analysis, GIS not only encourages the transfer of learning across a range of content areas, but it also reinforces the learning in each by demonstrating its applicability and relevance.

In the past 20 years we have witnessed a sea change in GIS, and those changes have profound implications for the classroom. Twenty years ago, GIS was a desktop-based software. To use it in classrooms, educators had to confront the often-difficult issues of software installation and data storage. Sharing and collaboration among students

Kathryn Keranen is a retired Fairfax County, Virginia, teacher and now an adjunct instructor at James Madison University who has taught geospatial technology to students and instructors for over 20 years. She coauthored (with Robert Kolvoord) the four-book series Making Spatial Decisions. *Lyn Malone is an educational consultant specializing in the classroom application of geospatial technologies and author of* Mapping Our World: GIS Lessons for Educators and the Teachers Guide to Community Geography.

Keranen and Malone are the coauthors of the Instructional Guide for the ArcGIS Book *from Esri Press.*

could be laborious. Today's educators and students can bypass those issues by using ArcGIS Online, enriching their own data with infinite open data resources, and fostering data analysis, sharing, and collaboration through the powerful capabilities of an ArcGIS organization. Further, the proliferation of data collection apps for mobile devices means that students can now take their GIS beyond the classroom walls to their real-world study areas.

GIS engages students in ways that no other teaching tool has ever done. With GIS, students have the power to address problems in their own communities and confront problems in which they have a personal interest. GIS enables them to be the agents of their own learning. As they collect, evaluate, and analyze data relevant to the questions they ask, the GIS workflow forces them to process that information and come to data-driven conclusions. As one teacher put it, "the student who makes the map is the student who learns."

The Partnership for 21st Century Skills, a public-private organization formed in 2002 to advance a new model of learning, asserts: "There remains a profound gap between the knowledge and skills most students learn in school and the knowledge and skills they need in typical 21st-century communities and workplaces." The integration of leading edge technologies (and GIS in particular) across the curriculum creates an essential bridge across that gap.

Here are three main thoughts. First, pay attention to your students. At least some part of your experience has to be observing and interpreting how your learners react to new software, and new curricula. Try to keep in mind that this community of students brings a whole new set of expectations and skills to the table. The first time we introduced *The ArcGIS Book: 10 Big Ideas about Applying Geography to Your World* to students, within 20 minutes a group of them opened the book's PDF version on their phone, leaving their laptop screen for their online mapping. We remember looking at each other, thinking, "Is this going work?" Yes, it did work, and in another five minutes, most of the other students were doing the same. Second, use what your students know to your advantage. Their skill with their mobile devices and apps means that you are not responsible for writing pages of instructions. Last, but most important, let your students ask *their* questions.

Quickstart

Drone mission checklist

According to Federal Aviation Administration (FAA) statistics, there were already more than 400,000 UAV (drone) pilots operating in the United States as of March 2016. Worldwide this number probably exceeds one million. So while this is an exciting time for drone hobbyists and professionals alike, it's important to understand the rules around drone flight in the nation. Other countries have their own rules (or not). Make sure you check with your local aviation authority before launching your mission. The following partial list is provided by 3DR, a leader in the citizen and commercial small drone space.

✔ Before leaving home, check that your controller, drone, and mobile devices are fully charged and ready to go. If you plan on using your GoPro®, make sure its battery is fully charged as well.

✔ In the field, check your flying location for people. Whether you are a veteran pilot or new to flying, never fly around people! Unforeseen flight hazards can occur at any time and you are ultimately responsible for the safety of yourself and others.

✔ Small drones (like the 3DR Solo) are not designed for harsh weather conditions. If it is windy, snowing, raining, or other inclement weather, pack up and wait for calmer skies.

✔ Stay at least five miles away from surrounding airports.

✔ Stay clear of crowded Wi-Fi environments. If you are in an area with many different Wi-Fi networks, your wireless connection to a drone will not be as reliable. Also, tall objects such as telephone poles and cell towers cause Wi-Fi interference, so make sure that you are clear of these as well.

✔ For a smooth takeoff, find a flat and solid launch surface. Make sure you have 20 feet of space in each direction to take off from and land on.

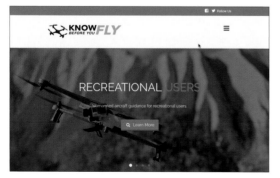

Know Before You Fly was founded by the two leading organizations with a stake in Unmanned Aircraft Systems (UAS) safety—the Association for Unmanned Vehicle Systems International (AUVSI) and the Academy of Model Aeronautics (AMA). The FAA is partnering with the founding members to spread the word about safe and responsible flying.

Learn ArcGIS Lesson

Get Started with Drone2Map

In this lesson, you'll run a complete inspection and mapping mission for an office building in Etterschlag, Germany. You'll download the trial version of Drone2Map and the sample data, and then use the software to process the data just as it looks as it comes off the drone. You'll set up the parameters that will create a number of 2D and 3D outputs, including an orthomosaic; 3D mesh; digital surface model; and even a special kind of sharing document, a 3D PDF. At the end, you'll publish your data tile service in ArcGIS Online.

▶ Overview

Drones provide an affordable means of capturing images of natural and man-made, land-based features and areas that may be difficult to access or fully cover because of size or terrain. Inspection of critical infrastructure using drones is a growing practice that improves the ability to inspect fixed assets, including those that span vast areas.

▶ Build skills in these areas:

- Working with Drone2Map
- Publishing 3D data

▶ What you need:

- Estimated time: 60 minutes
- Publisher role in an ArcGIS organization

Start Lesson

Esri.com/imagerybook/Chapter9_Lesson

Using high resolution sample data captured from this Solo camera-equipped quad-rotor drone from 3DR, you'll create a number of 2D and 3D GIS data products.

Contributors & acknowledgments

Contributors

Volume Editors: Clint Brown, Christian Harder
Series Editor: Christian Harder

Chapter 1—Clint Brown
Chapter 2—Christian Harder
Chapter 3—Steve Kopp, Melanie Harlow
Chapter 4—Gerry Kinn, Cody Benkelman
Chapter 5—Nawajish Noman,
            Vinay Viswambharan
Chapter 6—Brian Sims, Craig McCabe
Chapter 7—Kevin Butler, Daniel Siegel
Chapter 8—Mike Muller, Gerry Kinn,
            Peter Becker
Chapter 9—Christian Harder, Clint Brown,
            Tony Mason

Other contributors—Eric Rice, Greg Allord,
Jamie Ritchie, Kass Green, Kathryn Keranen,
Kathryn Sullivan, Lawrie Jordan, Lyn Malone,
Matt Artz, Mark Romero, Sarah Parcak

Acknowledgments

The work of a number of Esri cartographers
and data scientists is featured. Thanks to
Aileen Buckley, Allen Carroll, Andrew Skinner,
Jennifer Bell, Jim Herries, John Nelson,
Kenneth Field, Lisa Berry, Nathan Shephard,
Sean Breyer, and Richie Carmichael.

The Learn ArcGIS team is Bradley Wertman,
Brandy Perkins, John Berry, Kyle Bauer, and
Riley Peake.

Special thanks to Betty Martinez, Candace
Hogan, Catherine Ortiz, Chris Andrews,
Claudia Naber, Clayton Crawford, Cliff Crabbe,
Dan Hedges, Dan Zimble, Dave Boyles, Dawn
Wright, Eleanor Haire, Eric Laycock, Gisele
Mounzer, Jeff Liedtke, Jeremy Schuyler, Jie
Zhang, John Calkins, Kathleen Morgan, Kevin
J. Butler, Michelle McDaniel, Mike Kataoka,
Mike Livingston, Peng Gao, Renee Brandt,
Sandi Newman, Steve Pablo, Veronica Rojas,
and Wei Ying for project support, feedback,
and input.

Finally, thanks to the worldwide imagery and
GIS user community for doing amazing work
with ArcGIS technology.

Credits

38 The History Behind Flooding in Charleston, South Carolina: USDA FSA, DigitalGlobe, GeoEye, Microsoft, CNES/Airbus DS, NOAA Office for Coastal Management, Esri, HERE, DeLorme.

38 Shaded Relief Basemap: © 2014 Esri; Sources: Esri, DeLorme, USGS, NPS.

39 Photo by Christian Harder.

41 Landsat Shaded Basemap: Esri, United States Geological Survey (USGS), National Aeronautics and Space Administration (NASA).

41 Seasonal Charges: Earthstar Geographics | Esri, HERE, DeLorme

42 Orthorectified imagery: World Elevation Service, Esri, NASA, NGA, USGS.

43 Landsat 8 (Land/Water Boundary): City of New Orleans, Esri, HERE, DeLorme, USGS, NGA, EPA, USDA, NPS, NASA.

43 Landsat 8 (Near Infared, Bureau of Land Management, Esri, HERE, DeLorme, Intermap, USGS, NGA, EPA, USDA, NPS.

43 Landsat 8 (True Color): City of Huntsville, County of Montgomery, Esri, HERE, DeLorme, Intermap, USGS, NGA, EPA, USDA, NPS.

43 Landsat GLS (Vegetation Analysis, City of Huntsville, County of Montgomery, Esri, HERE, DeLorme, Intermap, USGS, NGA, EPA, USDA, NPS, NASA.

44 USA NAIP Imagery: VITA, Esri, HERE, DeLorme, increment p, Intermap, USGS, METI/NASA, EPA, USDA | Esri, USDA Farm Service Agency | VITA, Esri, HERE, DeLorme

44 USGS Historical Topographic Map Explorer: City of New Orleans, Esri, HERE, DeLorme, USGS, NGA, EPA, USDA, NPS.

44 The Scientific Data Discovery and Explorer: Esri, GEBCO, IHO-IOC GEBCO, NGS, DeLorme.

47 Arctic Ocean Basemap: Esri, DeLorme, GEBCO, NOAA NGDC, NOAA, National Geographic, HERE, Geonames.org, and other contributors.

47 Arctic Ocean Basemap (Sea Ice Extent): Esri, DeLorme, GEBCO, NOAA NGDC, National Snow and Ice Data Center, National Geographic, HERE, Geonames.org, and other contributors.

48 Jack Dangermond and Professor Dawn Wright—2012 Esri Oceans Summit: Esri.

49 Landsat Imagery/Access Image Services: Esri, USGS, Landsat NASA.

51, 52 Landsat Imagery/Access Image Services: Esri, USGS, Landsat NASA.

53 Momotombo Volcano: Esri, HERE, DeLorme, USGS, METI/NASA.

54 A World of Circles - Egmont National Park, New Zealand: Esri, Earthstar, CNES/Airbus DS.

54 World Imagery—Newfoundland: St. John's, DigitalGlobe, GeoEye | Esri, HERE, DeLorme, NRCan.

56 Landsat 8 Panchromatic & Pansharpened: IGN, Esri, HERE, DeLorme, USGS, NGA.

56 Landsat GLS (Agriculture): Esri, HERE, DeLorme, USGS, NGA, NASA.

56 The Moisture Index: How wet or dry?: Esri, HERE, DeLorme, FAO, NOAA, USGS.

57 Coastal Dynamics and Risks: Esri, HERE, DeLorme, NGA, USGS | ROLNP, 2014.

58 Monitoring severe foods: © CNES 2013-2014, Distribution Airbus DS.

58 Mining imagery for mineral patterns : Digital Globe; Images processed by MDA Information Systems LLC.

59 Copernicus: © CNES 2013-2014, Distribution Airbus DS.

60 Landsat 8 photo: NASA.

60 Landsat 8 photo: NASA.

60 Landsat Data Continuity Mission Overview: NASA.gov.

61 Sentinel-2 catches eye of algal storm: ESA.

62 Red Edge: MicaSense, City of Saskatoon, Bureau of Land Management, Esri, HERE, DeLorme, increment p, USGS, EPA, USDA, AAFC, NRCan.

64 KyFromAbove-Elevation: County of Montgomery, Esri, HERE, DeLorme, increment p, Intermap, USGS, METI/NASA, EPA, USDA | KYAPED Partners, DGI.

64 Oso Landslide 3D Swipe Before/After: Puget Sound Lidar Consortium, with imagery from ArcGIS Online. (Surfaces collected mid-2013), WSDOT, USGS.

65 Dynamique et risques côtiers: Esri, HERE, DeLorme, NGA, USGS | ROLNP, 2014.

65 NEXRAD Hawaii: Earthstar Geographics | Esri, HERE, DeLorme.

66 Thermal Infrared Imagery: MRLC, Bureau of Land Management, State of Oregon, State of Oregon DOT, State of Oregon GEO, Esri, HERE, DeLorme, INCREMENT P, Intermap, USGS, METI/NASA, EPA, USDA, US Forest Service.

66 Video #2—New Porter Ranch Gas Leak Timelapse (12/10/15): mmforthepeople.

67 Hyperspectral Signatures of Sand Samples: Esri; Donna Selch, Caiyun Zhang, Anton Oleinik, and Hannah Cooper, Florida Atlantic University.

68 3D view of 2-µm region vibrational mineral map draped on Google™ Earth topography looking NW toward Cuprite alteration centers (3x vertical exaggeration). AVIRIS image does not extend to mountains northwest of the Cuprite and Goldfield Hills areas so colors there are from Google™ Earth base images.

68 Sarah Parcak: Top photo by Louise Bray, BBC; bottom photo by Ryan Lash @ TED.

69 View from Drone2Map for ArcGIS: Data © 2015 Falcon Eye Drones. All Rights Reserved. Delair-Tech DT-18 Drones.

70 Esri's ArcGIS Full Motion Video Add-In Demonstrated at GEOINT 2015: GPS World.

71 Living Atlas of the World: Esri.

72 The Three Gorges Dam on the Yangtze River, China: Photo by Le Grand Portage under CC-by-2.0 license.

72 Poyang Lake shrinking for lingering drought: Photo by Xinhua/Zhou Ke © 2014 Xinhua, english.news.cn. All rights reserved.

73 Turning Imagery into Information: USGS.

75 The Cougar Corridors of Southern California: County of Los Angeles, Bureau of Land Management, Esri, HERE, DeLorme, Intermap, USGS, NGA, EPA, USDA, NPS.

76 Tree Canopy Land Cover: City and County of Honolulu, Esri, HERE, DeLorme, USGS, METI/NASA, NGA, EPA, USDA | DPP, Holis, University of Vermont Spatial Analysis Lab in cooperation with the USDA Forest Service Northeastern Research Station.

76 Red Edge: Micasense, City of Saskatoon, Bureau of Land Management, Esri, HERE, DeLorme, increment p, USGS, EPA, USDA, AAFC, NRCan.

76 Great Lakes Region Land Cover Report: Earthstar Geographics | NOAA Office for Coastal Management | Esri, HERE, DeLorme.

77 A River Reborn: Esri.

77 Cost Surface: Esri, HERE, DeLorme, USGS, NGA, EPA, USDA, NPS.

78 Explore a Tapestry of World Ecosystems: Esri, GEBCO, DeLorme, NaturalVue, Metzger et al., 2012, ESA, GEO, USGS, WorldClim.

79 Impervious Surface Analysis: County of Montgomery, Esri, HERE, DeLorme, increment p, Intermap, USGS, EPA, USDA.

79 Global Forest Watch Commodities: Esri, HERE, DeLorme, MapmyIndia, © OpenStreetMap contributors, and the GIS user community.

80 How to Use Land Cover Data as a Water Quality Indicator: Earthstar Geographics | Department of Commerce (DOC), National Oceanic and Atmospheric Administration (NOAA), National Ocean Service (NOS), Office for Coastal Management (OCM) | US Census Bureau | Esri, HERE, DeLorme.

80 MN Solar Suitablility: mn.gov/solarapp.

81 Swift Wind Farm Viewshed Analysis: DoBH, OS, Esri, HERE, DeLorme, USGS, NGA.

81 Decisive Moments in the Battle of Gettysburg: Esri, HERE, DeLorme, increment p, Intermap, USGS, METI/NASA, EPA, USDA.

82 Modeling Flash Flood Events in Ungaged Semi-Arid Basins, USGS
The National Map: National Boundaries Dataset, National Elevation Dataset, Geographic Names Information System, National Hydrography Dataset, National Land Cover Database, National Structures Dataset, and National Transportation Dataset; US Census Bureau—TIGER/Line; HERE Road Data.

82 USA Soil Flooding Frequency: Esri, HERE, DeLorme, NGA, USGS | Esri, HERE | Source: USDA NRCS.

83 Soils of the Panoche Hills: Esri, HERE, DeLorme, increment p, NGA, USGS, USDA - NRCS, NGA, NASA, CGIAR, NCEAS, NLS, OS, NMA, Geodatastyrelsen,GSA, and the GIS user community.

85 Global Precipitation Runoff: Esri, HERE, DeLorme, FAO, NOAA, NASA.

85 Pacific Ocean Science and Exploration: Esri.

86 NC 12 at the north end of Rodanthe on Hatteras Island, shown here on Sunday, Aug. 28, 2011, was washed out by Hurricane Irene. Steve Earley | The Virginian-Pilot.

86 011 NOAA National Geodetic Survey (NGS) LIDAR of North Carolina: Hurricane Irene Emergency Response Mapping Product: NOAA's Ocean Service, Coastal Services Center (CSC).

87, 89 Charlotte, North Carolina, Impervious Surfaces: Jefferson County (KY) Property Valuation Administrator.

92 Lidar Point Cloud with Textured Buildings of San Francisco: Esri, EagleView Pictometry, NOAA.

93 Visualize New Developments: MRLC, Oregon Metro, Bureau of Land Man-agement, State of Oregon, State of Oregon DOT, State of Oregon GEO, Esri, HERE, DeLorme, increment p, Intermap, USGS, EPA, USDA, US Forest Service.

95 3D Across All Scales: Esri.

96   Global scale: Esri, NASA.
96   Esri campus 3D: City of Redlands, County of Riverside, Bureau of Land Management, Esri, HERE, DeLorme, increment p, Intermap, METI/NASA, USGS, EPA, USDA.
96   Montreal, Canada, Scene: Esri, USDA FSA, DigitalGlobe, GeoEye, Microsoft, CNES/Airbus DS.
96   Esri campus 3D: City of Redlands, County of Riverside, Bureau of Land Management, Esri, HERE, DeLorme, increment p, Intermap, METI/NASA, USGS, EPA, USDA.
97   Point Cloud of the Grand Coulee Dam Symbolized by LAS Class Code: Esri, NOAA Coastal Services Center.
97   Detailed lidar collected for parts of Petaluma, California, by the GIS team in Sonoma County: NOAA.
100   My Topo Scene: Esri, HERE, DeLorme, USGS, GA.
100   My Scene imagery: Esri, Earthstar Geographics, CNES/Airbus DS.
101   Separation of vegetation biomass from the bare earth surface: Fugro Geospatial, EarthData.
101   Drone2Map comparison: Data © 2015 Falcon Eye Drones. All Rights Reserved. Delair-Tech DT-18 Drones.
102   Landsat Natural Color Imagery: Esri USGS, NASA, Landsat.
102   National Geographic World Map: National Geographic, Esri, DeLorme, HERE, UNEP-WCMC, USGS, NASA, ESA, METI, NRCnn, GEBCO, NOAA, INCREMENT P.
102   3D Buildings Model: Esri, Pictometry International.
103   Building footprints: Esri.
103   3D Tree Model: Esri.
107   Photo textured realistic Mont Blanc in the French Alps: Esri France, SRTM, NASA, Landsat, USGS.
107   Pictometry 3D Scene with Indianapolis and Philadelphia 3D Scene layers: Esri, Pictometry International.
113   Katrina at 10: NOAA, Department of Commerce.
116   Seasonal Charges: Earthstar Geographics | Esri, HERE, DeLorme.
116   Decisive Moments in the Battle of Gettysburg: Esri, HERE, DeLorme, increment p, Intermap, USGS, METI/NASA, EPA, USDA.
116   Drought Tracker: Esri, HERE, FAO, NOAA | © 2015 National Drought Mitigation Center | © 2013 Esri.
117   Visualizing Change Over Dubai: Farsi Geotech, HERE, DeLorme, USGS, METI/NASA, NGA.
117   Mile 216.3, Confluence of Green and Colorado Rivers, Upstream View from River Right (Stake 2429), 31 May 1889. Photo by Franklin A. Nims, 57RS35, courtesy of The National Archives.
117   Mile 216.3, Confluence of Green and Colorado Rivers, Upstream View from River Right (Stake 2429), 24 March 1997: Photo by Robert H. Webb, courtesy of USGS.
118   Hurricane Katrina: NOAA (video).
118   GLDAS Soil Moisture-Copy: Esri, Earthstar Geographics, NASA.
118   Katrinahd1080.mp4: NOAA.
119   Human activities are reshaping Earth's surface: Esri's World Imagery Basemap, 1990 Landsat Imagery.
119   Spyglass on the Past: USDA FSA, Microsoft | David Rumsey Historical Map Collection.
122   David Rumsey Historical Map Collection: www.davidrumsey.com.
122   Africa 1787: David Rumsey Historical Map Collection.
122   Pittsburgh Historic Maps: University of Pittsburgh.
122   1942 Theater of War in Europe, Africa, and Western Asia: Esri, © National Geographic Society.
123   USGS Historical Topographic Map Explorer: City of New Orleans, Esri, HERE, DeLorme, USGS, NGA, EPA, USDA, NPS.
124   Historical Netherlands maps: Tijdreis.
125   Post-Tornado Imagery of Oklahoma City and Moore Taken May 22, 2013: Texas Parks & Wildlife, Esri, HERE, DeLorme, Intermap, USGS, NGA, EPA, USDA, NPS | Bearing Tree Land Surveying (http://www.btls.us/), Esri, State of Oklahoma.
125   USA Wildland Fire Potential: Esri, HERE, DeLorme, FAO, NOAA, USGS, EPA | Source: Fire Modeling Institute (FMI), USDA Forest Service, Esri.
126   National Weather Service Precipitation Forecast: Esri, HERE, DeLorme, NGA, USGS.
126   Before and After the 2015 Tornadoes in Illinois: Microsoft | National Oceanic and Atmospheric Administration (NOAA), National Ocean Service (NOS), National Geodetic Survey (NGS), Remote Sensing Division | Esri, HERE, DeLorme, iPC.
126   USA Precipitation Forecast: Esri, HERE, DeLorme, NGA, USGS.
126   Katrina + 10: A Decade of Change in New Orleans: Esri, HERE, DeLorme, NGA, USGS.
127   Living Atlas of the World: Esri.
128   Atlas for a Changing Planet: Esri.
128   Tracking Land Cover Change in Connecticut: University of Connecticut, Center for Land Use Education and Research (CLEAR).
128   The King Tides Project: Esri, NOAA, Earthstar Geographics.
129   72-hour precipitation from the US National Weather Service: Esri, HERE, DeLorme, NGA, USGS.
129   World Temperature Change 2050 Scenario 8.5: Esri, GEBCO, DeLorme, NaturalVue, HERE, IPCC.
130   Sea Level Rise and Coastal Flooding Impacts: NOAA National Ocean Service.
131   Devastation in Nepal: Kathmandu Before and After the April 2015 Earthquake: Esri, CNES/Airbus DS | Airbus Defense and Space.
131   The Assassination of Abraham Lincoln: Esri.
131   Outlook for European cities: Higher temperatures and longer heat waves: Esri.
132   Samut Songkhram, Thailand Historical Change: Esri, HERE, DeLorme, USGS, NGA.
133   A map of a copper mine near Salt Lake City. A GIS is built on the concept of layers: a base map layer, a soil layer, a climate layer, and so on: Esri.
135   Landsat 7 Acquisition Calendar: USGS.
136   NASA Earth Observing Fleet: NASA/Goddard Space Flight Center Scientific Visualization Studio.
137   Landsat App: City of Riverside, County of Riverside, Esri, HERE, DeLorme, NGA, USGS.
138   Landsat 8—Points of Interest: Esri, HERE, DeLorme, FAO, NOAA.
140   Massachusetts 2013-2014 USGS Color Ortho Imagery: USGS, MassGIS, Esri.
140   Global Land Cover Facility: University of Maryland, NASA, GOFC-GOLD.
141   Landsat Polar App: Earthstar Geographics, Esri, USGS, NASA.
142   Landsat Global Land Survey: Esri, USGS, NASA Landsat.
143   Landsat Global Land Survey: Esri, USGS, NASA Landsat.
144   Bringing Science to Life, Dr. Kathryn Sullivan (video): Esri.
144   Photo by NOAA.
147   Lidar-Raw Data: OpenTopography as a subset of Sonoma County dataset.
147   Lidar-Color Classified: OpenTopography as a subset of Sonoma County dataset.
147   Lidar-Points Assigned RGB Values: OpenTopography as a subset of Sonoma County dataset.
147   Full Motion Video Add-in: Esri.
149   Download Imagery from an Online Database: Esri, USGS, NASA Landsat.
153   LA Hub: A Blueprint: Esri.
153   National Map website: Australian Government.
155   Berlin DOP20: Esri, DigitalGlobe, GeoEye, Earthstar Geographics, CNES/Airbus DS, USDA, USGS, AEX, Getmapping, AeroGRID, IGN, IGP, swisstopo | Geoportal Berlin, DigitalGlobe, Microsoft | iPC.
156   Orthomosaic Image Maps Created with Drone2Map and Solo-Copy: City of Olympia, Bureau of Land Management, Esri, HERE, DeLorme, increment p, Intermap, USGS, EPA, USDA.
157   Landsat Layers: Esri, HERE, DeLorme, NGA, USGS.
158   3D GPS Activity: 33-km, 1:31-minute Bike Ride: Esri, USDA, FSA, Microsoft.
158   First time with the Oculus Rift (Esri CityEngine SIGGRAPH 13, Autobrine).
159   Comparing DSM with Orthophoto Imagery: Credit: DigitalGlobe, Microsoft | Salzburger Geographisches Informationssystem (SAGIS).
159   Orthomosaic Image Maps Created with Drone2Map and 3DR Solo: City of Olympia, Bureau of Land Management, Esri, HERE, DeLorme, increment p, Intermap, USGS, EPA, USDA.
163   Kathryn Keranen and Lyn Malone: Photo courtesy Keranen and Malone.
164   Solo Drone: 3DR.

The deep purple in the lower right spreads out into a few channels before fading into a multitude of colors. These channels are remnants of an ancient drainage network in Kenya. The beauty of the colors actually hides a stark reality for hundreds of thousands of people. The dark spots at the top center of the image are refugee camps.